PRAISE FOR CHARLES LONDON AND
One Day the Soldiers Came

"Charles London's remarkable writing and sensitivity takes us on a profound and deeply moving journey into the minds of children who live with war: the memories of it, the loss of families, and the displacement of their identities and cultures. But, most important, he shows us how these children maintain their humanity even when hope seems lost. I thank him for this."

—Ishmael Beah, author of *A Long Way Gone*

"Charles London has somehow managed to write a book about children in war that is not sentimental, not maudlin, not shrill, and certainly not preachy. Instead, *One Day the Soldiers Came* is an unblinking account of a peculiar human reality that extends even down to the littlest of us—namely that we, adults and children alike, are a species of relentless survivors, equally capable of committing unspeakable horrors or noble acts of altruism, sometimes both on the same day, in order to endure. It must have been a difficult task to face down such contradictions in an eight-year-old child, for instance, but London has done so in a wise and captivating story."   —Elizabeth Gilbert, author of *Eat, Pray, Love*

"By taking us into the world of innocent children torn apart by war, Charles London brings an uncomfortable truth to life. This book is often difficult reading, but attention must be paid."

—Ambassador Richard Holbrooke

"By serving as our eyes, ears, and legs in some of the most troubled places in the world, Charles London brings us into realms of experience that we would prefer, in our weaker moments, to ignore. The stories told by these children, however difficult they may be for us to read, are essential to our understanding of the world today. And essential to our humanity as well; more than once during the reading of this book, I was compelled to get up from my chair, walk down the hall, and give my children a hug."

—Ben Fountain, author of *Brief Encounters with Che Guevara*

"An eye-opening exposé on the permanently scarred children of war in 'situations where violence and hardship are the norm.' . . . Thankfully, London offers positive updates on several of the youth. But readers will never forget the physical and mental damage inflicted upon their developing minds. Neither sentimental nor patronizing, these are harsh, numbing experiences. Searing and heartbreaking."

—*Kirkus Reviews*

ONE DAY THE SOLDIERS CAME

Voices *of*
Children *in*
War

CHARLES LONDON

HARPER PERENNIAL

NEW YORK • LONDON • TORONTO • SYDNEY

HARPER PERENNIAL

P.S. ™ is a trademark of HarperCollins Publishers.

ONE DAY THE SOLDIERS CAME. Copyright © 2007 by Charles
London. All rights reserved. Printed in the United States of America.
No part of this book may be used or reproduced in any manner what-
soever without written permission except in the case of brief quota-
tions embodied in critical articles and reviews. For information address
HarperCollins Publishers, 10 East 53rd Street, New York, NY 10022.

HarperCollins books may be purchased for educational, business, or
sales promotional use. For information please write: Special Markets
Department, HarperCollins Publishers, 10 East 53rd Street,
New York, NY 10022.

FIRST EDITION

Maps by Amy Barrett

Designed by Nicola Ferguson

Library of Congress Cataloging-in-Publication Data is available upon
request.

ISBN: 978-0-06-124047-8
ISBN-10: 0-06-124047-8

07 08 09 10 11 OV/RRD 10 9 8 7 6 5 4 3 2 1

To all the parents, mine and theirs, alive or dead, who try, against the odds, to protect us.

That girls are raped, that two boys knife a third,
Were axioms to him, who'd never heard
Of any world where promises were kept,
Or where one could weep because another wept.
—from W. H. Auden's Shield of Achilles

We that are young
Shall never see so much, nor live so long.
—from William Shakespeare's King Lear

When Elephants fight, it is the Grass that suffers.
—East African Proverb

CONTENTS

FOREWORD

ROBERT COLES, M.D.

During the late 1950s I worked as a resident in pediatrics and child psychiatry on the wards of the Children's Hospital Boston; and so doing, I met many children who were not only sick but hurting with unremitting pain, debilitating to both mind and heart—to the point that some boys and girls dared say to their parents (and to us doctors), that they wished for death, whose arrival, they averred, would end the agony they no longer felt able to bear, even with a modicum of equanimity. One day, as I talked with a ten-year-old lad who had contracted polio, and who was paralyzed from the waist down (no vaccine was available then, to spare children from such a dreadful, disabling disease), I heard this from Jimmie: "My dad fought in the war [the Second World War], and he said he saw a lot of kids my age get killed—and he even saw some fighting hard, 'like good soldiers, so their country [France] could be free,' from that dictator, Hitler, and his army." A moment of silence, and then, as if my perplexity had become quite apparent, this soliloquy of sorts: "You have to be brave, and keep on fighting. If kids could fight for their country in Europe, I sure can try to fight for myself, right here and now. 'Be a good soldier,' my mother and dad tell me, and then I say, 'You bet I will.' So, when they come to visit me, they ask how the soldier is doing, and I say, 'The soldier is fighting hard, and I hope he wins the war.'"

There are, of course, soldiers and *soldiers*; indeed, young Jimmie, before his hospitalization, had often played soldier games in the backyard of his suburban Boston home. He and his friends had taken sides, shouted and screamed at one another, aimed sticks as if they were guns and made noises—*bang, bang, bang!*—to affirm deadly intent. "My dad was a soldier, and me and my pals fight like soldiers, and our dads coach us," Jimmie once told me. Yes, indeed, here were American children and their parents (one-time warriors in Europe and Asia), engaged in vigorous military activity, so it seemed to all who watched: "My mom," Jimmie told me, "said I could go to join the army, and they'd not have to teach me much, because of what my dad and the other dads [of his neighborhood friends] have taught me—and remember, in a real war, kids sometimes fight too, or they sure see the fighting right before their eyes."

That long-ago critical moment in my occupational life came back to my mind as I read the pages that follow—their collective words an unforgettable lesson for all of us readers: children become witnesses of war fought, and further, children become warriors themselves, ready and willing to take up arms, even as they observe others doing likewise—a violent world registering its implacable philosophy on others, who are violated in the name of this or that slogan, creed, military or political or social or national reality. "You know in battle, a lot of times it's the blind leading the blind, the good fighting the bad, the smarties foxing out the dummies, the lucky ones beating out the ones who are dying, down on their luck"—so Jimmie had learned from his veteran soldier dad, and from other dads (those of his friends), who had also fought for America in the 1940s.

Now, we readers of this book can meet children like Jimmie and his next door pals—young ones not playing war games for fun (or at the behest of remembering parents, alive and doing well in the peaceable kingdom of the United States of America), but, alas, swept into ongoing warfare, and become, willynilly, actual combatants or victims of others wielding guns, knives, and bombs.

Ahead are those children, and ahead for us who meet them through a book's knowing, resolute insistence, is plenty to ponder: knowledge offered becomes ours to have, to hold up to our minds' eyes for sustained consideration. We are offered, too, in this volume's extraordinarily affecting presentation, the valuable words of Anna Freud; and as I met them, I kept on remembering her thoughts about children, caught in the turmoil of the war being waged near their homes, their families, and their friends. I was privileged to know her, hear her recall the past, and reflect upon what she had learned (so often) by watching boys and girls attentively, and keeping in mind what they had said to her.

"So often, children learn violence from others," she once remarked to me, and then this follow up: "That is obvious [what she had said], but not so obvious, at least to some of us who worked with children in London, during the time of the [Nazi] blitz was the lure, you could call it, of violence, of war, of aggression visited, and then returned in kind. My father in his writings knew to emphasize aggression as an aspect of all of our psychological lives, but when that 'drive,' he called it, becomes the norm, so to speak—and the young are summoned to what might be termed 'a call to arms'—then, in a sense, aggression is given the sanction of the adult world, and enacted by it. Here we have, under such circumstances, an extraordinary kind of childhood being allowed (encouraged even) by parents, teachers, and civic authorities: boys and girls prodded (taught even) by adults to be fighters, to join with adults in their attitudes, feelings, and, yes, their actions. I saw in London some children fighting as if they were 'the Germans' or 'on our side'—and I knew that in Europe, during the war, some children fought alongside adults, as ones who did errands, surely, but also as ones not only spying, running here and there, but taking orders—the young become fighters alongside their elders."

A moment of silence, then this: "So it goes, children become fighters, warriors"—and today, thanks to this compelling book, the rest of us can know, as Anna Freud put it, how "it goes," in our twenty-first century for children across the globe, caught in the throes of war, become witnesses to it, become soldiers in it—struggling for victory over others, and all the while, struggling to grow up in an all too callous, even murderous, world.

AUTHOR'S NOTE

During the spring of my junior year of college, when I was twenty-one years old, I began this project, collaborating with Refugees International as a Research Associate. Over the next five years, I traveled to eight countries, spending up to a month at a time interviewing children and their caretakers, visiting their homes and schools, playing soccer, and doing drawings. The world changed a great deal in that time and the children's attitudes towards me, an American, shifted somewhat as well (which is why I did not visit Iraq or Afghanistan). The following timeline gives a sense of the major events in this book.

- June 28, 1389: The Battle of Kosovo brings the kingdom of Serbia under Ottoman control.
- March 1962: General Ne Win leads a military coup in Burma.
- September 1983: The Sudanese government triggers the Second Sudanese Civil War by imposing Islamic Law on the Christian and Animist south of the country. The war displaces nearly four million Sudanese, including thousands of children.
- November 1965: Joseph Desiré Mobutu overthrows

the president of the Congo and renames the country Zaire.

- April 1992: The Bosnian Parliament, following the lead of Slovenia and Croatia, passes a referendum declaring Bosnia's independence from Yugoslavia which triggers a civil war supported by the Serb controlled government in Belgrade.

- April 1994: Rwandan President Juvénal Habyarimana is assassinated and the act is immediately blamed on Tutsi extremists. Under the pretext of national security, the Hutu Power government of Rwanda begins a mass extermination campaign to kill every Tutsi in the country. The international community does nothing to stop the killing. Paul Kagame, a Tutsi general, takes over Rwanda in July 1994 and puts a stop to the killing, sending hundreds of thousands of Hutus, including those responsible for the genocide, fleeing into Zaire.

- November 1996–May 1997: The First Congo War erupts and Laurent Kabila, with the support of Paul Kagame, overthrows Mobutu, ending his thirty-year reign.

- July 1998, Laurent Kabila severs ties with his Rwandan backers. Ethnic Tutsis in the eastern Congo rebel against his government with the support of Rwanda and the Second Congo War begins.

- March 1999: NATO bombardment of Serb positions ends the ethnic cleansing of the Kosovar Albanian population in the Serbian province of Kosovo. When Serb forces withdraw, the Kosovo Liberation

Army begins exacting revenge on the Serb civilians left behind. The United Nations takes over administration of the province.

- May 2001: I begin research for this project.
- July 2001: I spend a month in the Burundian and Congolese refugee camps in western Tanzania.
- September 11, 2001: Terrorist attacks on the Pentagon and the World Trade Center kill over 3,000 people. Images of the burning towers are seen even by children in the jungles of Thailand.
- October 7, 2001: The United States begins Operation Enduring Freedom to overthrow the Taliban in Afghanistan. Ensuring girls' access to education is one of the stated goals of the new government.
- January 2002: I visit the war-ravaged eastern Congo, passing through Rwanda.
- January 17, 2002: Mount Nyiragongo erupts in the Congo, forcing the evacuation of half a million people, including the research team from Refugees International, who retreat back to Rwanda.
- September 2002: I visit Thailand to research the lives of refugees and illegal migrants living in Bangkok and on the Thailand-Burma border.
- March 2003: The United States goes to war in Iraq. The pre-war planning does not include cogent provisions for children in the aftermath of the regime; many young people join the insurgency.
- May 2003: I visit Kakuma Refugee Camp, near the Kenya-Sudan border to interview the Lost Girls

of Sudan. At the same time, the United States resumes its resettlement program of the Lost Boys and a large number of Somali Bantus from Kakuma. The program had been suspended due to the 9/11 attacks.

- July 2003: A transitional government takes control of the Democratic Republic of Congo, officially ending the civil war, though violence in the east continues.

- March 2004: Rioting erupts across Kosovo when three children drown under mysterious circumstances.

- June 2004: I visit Kosovo to research the lives of refugee children returning home and trying to rebuild their lives amid ethnic tensions. I also spend a month in Bosnia with young adults who had been children during the war ten years earlier to learn how they are coping with peace.

- November 2004: Rebecca, a Sudanese "Lost Girl" is reunited with her cousin in the United States.

- October 2005: A small group of Congolese refugees in a Tanzanian refugee camp begin the repatriation process, returning to the eastern Congo. Over the next year, thousands more follow. I have no idea if the children I met, now young adults, are among the refugees returning home.

- July 30, 2006: The first free elections since independence from colonial rule occur in the Democratic Republic of Congo. Their outcome is immediately contested. Many of the children I met in the refugee camps in 2001 are now old enough to vote.

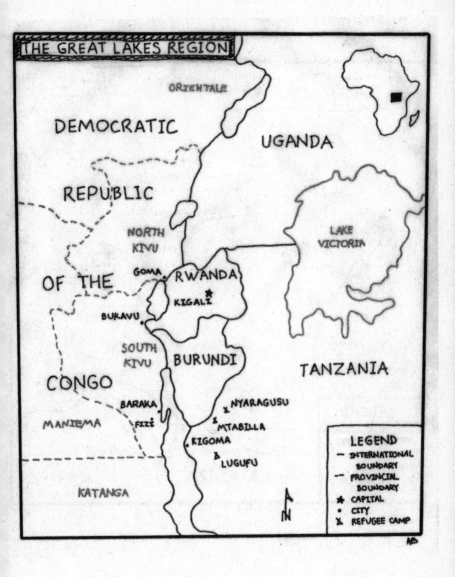

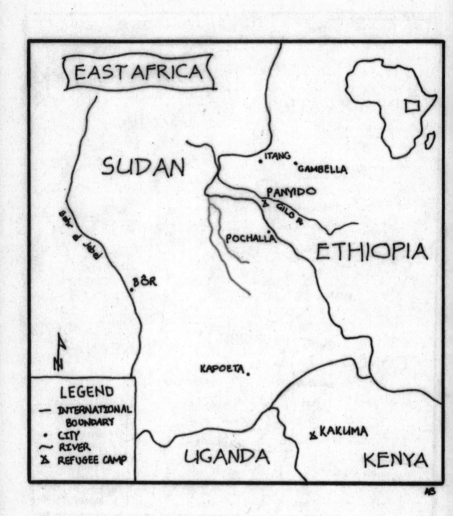

THE BALKANS

CROATIA

BELGRADE ★

BOSNIA &
HERZEGOVINA

SERBIA

SARAJEVO ★

FEDERATION of
BOSNIA and
HERZEGOVINA

REPUBLIKA
SRPSKA

RIVER IBAR

MONTENEGRO

MITROVICA

PEC   PRISTINA

GRANCANICA

KOSOVO

A
N

LEGEND
── INTERNATIONAL
   BOUNDARY
─ ─ PROVINCIAL
   BOUNDARY
─·─ DAYTON ACCORD
   BOUNDARY
   (EST. 1995)
★ CAPITAL
• CITY

ALBANIA

ADRIATIC
SEA

MACEDONIA

4B

ONE

"Innocent in the Ways of the World"

Childhood and War

It happened to Keto. He was sitting in school with his brother listening to the teacher recite their French lesson. *Je m'appelle, tu t'appelle, il s'appelle. . . .*

It happened to Michael. He was at home with his mother and father. He sat in the back room doing whatever it is that teenage boys do in back rooms, daydreaming, making plans, goofing off.

It happened to Nora on a sunny day. She was playing in the front of her house.

Patience and Charity were too young to remember what they were doing when it happened.

It happened to Nicholas, as it happened to the others. To his entire village, one day, it happened.

The soldiers came.

"It was a sunny day," Nora says, as if the weather were the most amazing thing. How could it happen on a sunny day?

"They put a knife on my neck," she says, the little blonde.

Picture her at eight years old, smiling and playing in the yard on a sunny day. They put a knife to her neck. "They wanted to rob us and they saw my mother's wedding ring and they told her to give it to them," she says. But the ring was hard to get off. Her mother struggled with it. The soldiers laughed. "Hurry up or we'll just cut off your finger!" they shouted. Her hands shaking, she got the ring off and they let her daughter go.

They shot Nora's uncle, though.

"They shot him with a silencer and then wrapped him in a carpet so his body would burn more easily." It happened on a sunny day in the Balkans when she was eight years old. When she was playing outside.

Keto and his brother saw people running, cattle running, the entire village near Baraka in the eastern Congo on the move. They heard gunfire. The teacher told them to go home; it was time to flee. Keto ran, clutching his schoolbooks to his chest. Barely four feet tall, charging through hell.

"The Mayi Mayi were yelling 'fire, fire,' commanding the village to be burned," Keto says. Flames tore at the thatched roofs of houses. People ducked low and tried not to catch the fury of the soldiers. They were looting the marketplace.

"When I went home, I didn't find my parents. My brother and I didn't know where my parents or grandparents were." They stood for a while in their empty home, calling for anyone they knew. With gunfire and flames around them, the two boys decided they must escape on their own. They made their way to the lake still clutching their schoolbooks to their chests. "They were our only possessions when we fled. I still have them." He nods, proud that he had held on to his books all these years, through such a long journey, after so many people have died.

Nicholas doesn't like to talk about what happened. He's thirteen years old, originally from Burma, though exiled in

Thailand now. He has seen crucifixions, executions, abductions.

Michael has seen the damage a machete can do to human flesh, his mother's flesh, his father's flesh.

Patience, from southern Sudan, has been raped, repeatedly.

They all draw pictures. Whether they like to talk or not, they jump at the crayons and markers, remake their world on paper. Their visions are at turns dark and painful, others are hopeful, light-hearted, nostalgic. It depends on the child, depends on the day. They all draw. We draw together. It's one of two activities we do. We draw and we play soccer. It is with soccer that everything begins.

You cannot know the children of a world at war until you begin to play soccer. You can interview them, as many have done, as I have spent countless hours doing, and you can read reports and studies and you can watch them do all manner of things and you can hear and hear and hear about children in war from just about everyone: charities and warlords, generals and social workers, parents and doctors and politicians. Everyone likes to talk about them. Children are the canvas on which societies paint themselves; their hopes, their hates and fears, their nationalist fantasies, their impossible dreams. The rhetoric is everywhere. Save the Children and Islamic Jihad use images of children in terribly similar PR campaigns.

Who are these children, though? The ones we see on the news, all wide-eyed and suffering in refugee camps. The ones we see in magazines, dressed in camouflage, firing rifles taller than themselves. The ones overflowing in history, the anonymous displaced, disenfranchised, photographed but not named, talked about but not remembered? Who are they?

Play soccer with them and you'll know.

* * *

The Eastern Democratic Republic of Congo, January 2002.

I'm in a center for demobilized child soldiers run by a non-profit organization in the city of Bukavu. Built on the coast of Lake Kivu, the city rests like a blanket on the hills. Dilapidated colonial buildings command views of lake sunsets and jungle horizons. In the morning you can hear fishermen singing on the lake as the sun comes up. The smell of wood fires fills the air, cooking goat meat and ground cassava. The wilderness around the city is as stunning as it is dangerous. Stories of massacres and banditry trickle in from the outlying areas. When it is quiet, you might hear the mist crackle with gunfire. Children from all over the eastern Congo seek safety from the fighting and some kind of livelihood on the muddy and crumbling streets. Some estimates put the number of homeless children in the city at twenty thousand.

I play soccer in a grubby courtyard belonging to a local charity. It's not much of a playground, filled as it is with giant puddles, pits, and loose shrubs, fenced in on all sides from the street, but the children have each other and an ingenious ball made of bundled plastic bags and string. That's enough for them. The game is informal. To an outsider, it looks like a skirmish, all shuffling feet and half-playful shoulder shoves. There are rules, rituals, a code of behavior, but none I'll be privy to. You could watch these children play for years and never see the current underneath, the history that creates this game, that's passed it on through generations of kids just like these. They don't need to learn the rules, they don't exist in words. They're in the blood. In the birth. There are no goal posts because there are no goals. Scoring is not the objective here, nor winning. The play is for the sake of play. Goals and points are finite, they imply a beginning and ending. This game has no beginning. It started long ago with other children and goes on in barrios and slums and ghettos and camps and shantytowns the world over. It will never end.

I'm no good at soccer, and the ball passes through my legs. I twist to stop it, putting my left foot behind my right as I step backwards, tripping myself into the mud. Pratfall. The children with whom I'm playing burst into laughter. The oldest among them is sixteen; the youngest is ten. All of them are trained killers.

The rebel group Rasemblement Congolaise pour la Democratie (RCD-Goma) controls this area, though it is often seized by paroxysms of violence from other factions or ethnic militias. The children have seen combat in a war widely known to be fought against civilians. I'm here to learn from them, to hear their stories, but we haven't gotten that far. Now it's time to play. Later, they'll tell me about their killings, their wars gone by, their nightmares, and their hopes. But not yet.

They've fought in different armies and come from different parts of the country. Fate has thrown them into this center together, turned them into a group, labeled child soldiers or ex-combatants or in some documents "youth who participate in armed conflict." The labels tell you little. In the language of humanitarian aid, there are many categories for children: Street Children, Internally Displaced Children, Child Soldiers, Child Heads of Household, Unaccompanied Minors, Children in Conflict with the Law, Children Affected by HIV, Children Accused of Sorcery. Categorization is a way of processing children for targeted assistance in crisis situations. Most children will fit into more than one category; few children in a war zone will fit into none.

No help there. Want to know them? Play soccer.

Xavier plays delicately at first, kicking the ball as though it were made of glass. He lets the others charge in, lets them do the slide tackles. When the ball goes wide, he'll chase it and bring it back toward the group. They cluster in front of him as he approaches, all defenders, no allies in attack. He charges right in, daring the whole pack of them , trying to come through the other side in control of the ball. Now he throws elbows, now the grunts and shoulder-shoves. That's the game. You get the ball and you try to keep it. Xavier gets the ball and goes into the pack, goes looking for trouble. One doesn't avoid trouble in this game; safety is not the goal. Risk, that's the game. Get the ball and hazard losing it the moment it's yours. It's a game of constant loss.

I shudder with a thought about little Xavier, who must be about fourteen years old, who plays soccer in flip-flops, whose skinny legs poke out from his ragged shorts, whose Adidas T-shirt is torn and far too big for him. I wonder how many people he has killed.

Paul is always in the fray. He's about four feet tall with big brown eyes. He has the looks of a boy who could play the cute little brother with the snappy comebacks on any sitcom. He shoves like the rest of them, but he smiles widely, his teeth glowing white (these children have no access to dentists, but they have no access to candy either). When others fall, he helps them up. He makes sure everyone gets to play. He's thoughtful of me, trying to make sure I get the ball from time to time, trying to make sure I obey the amorphous rules—sometimes touching the ball means the game stops and you surrender it, sometimes you touch the ball and whoever kicked it has to be the monkey in the middle, sometimes it means nothing and the game just goes on and on and on. Paul is an excellent guide, a first-class soccer mentor for me. How did he end up here and not in school or on the set of a sitcom practicing his smile for adoring fans? What does he see when he closes his eyes at night? What does he hear?

* * *

Another soccer game, a continent away.

I nearly kick a soccer ball into a passing NATO truck with a machine gun mount. The gunmetal shines in the sunlight. The flag on the side suggests it's the Swedes. We hold our breath as the ball soars toward the armored vehicle. It arcs over the grass, bounces once and careens into the road. It misses the truck by a few feet and the soldiers keep driving. The truck leaves a trail of dust that takes half an hour to settle. The Serb elementary school students around me laugh and sigh with relief, and the oldest among them, Marko, twelve, runs off after the ball with a roll of his eyes. The kids had been chasing my missed kicks all afternoon. It's a steaming July day. I don't just sweat. I lose water in buckets. Kosovo, 2004. The children all ask me the same question.

"Do you know the Battle of Kosovo?"

Marko was the first of these children to ask it, perhaps because he was the biggest, the most handsome, the best soccer player (except perhaps for Katja, who is surgical with her kicks and dazzling with her footwork, but she's a girl, so doesn't count, as the boys explained to me privately).

"I saw Kosovo Polje on my drive here," I answered, citing the field where the battle took place. The field was speckled with daisies and buffeted with high voltage power lines. It did not strike me as a likely place for events of great magnitude. It could have been any number of fields dotting Kosovo. It looked, in fact, like any roadside field in middle America. Flat and slightly sun-scorched, stuck between two major routes, north to Mitrovica and west to Peja (Peč in Serbian—the names matter). The only remarkable thing about this field was that everyone with whom I spoke brought it up. When I asked about the conflict between Albanians and Serbs, they would say, like Marko, "It's the history. Do you know the Battle of Kosovo?" Over and over again, this refrain, "Do you know the Battle of Kosovo?"

The field saw a lot of bloodshed, terrible violence between Muslims and Christians, with casualties on both sides. The small province of Kosovo still reels from the battle. The leaders of both armies died in the conflict. The battle on Kosovo Polje secured Slobodan Milošević's power over the crumbling Yugoslav state in the late eighties. Serb children still draw pictures of it, lamenting Serbia's loss, the cause of all their present woes. The myth of the battle, the myriad interpretations of the story, of the massacres and war crimes, could easily hurtle the province of Kosovo back into civil war. This is remarkable because the Battle of Kosovo was fought on June 28, 1389.

The soccer game stops. Play makes room for history, and the children begin to tell the story. Marko went to the bench near our patch of field and grabbed a drawing from the table. We'd been drawing pictures before the soccer ball came out and hadn't had time to talk about them. The drawing belonged to Miroslaw. He was the littlest one in the group and better only than me at soccer. The pause in play must have been a relief to him. He reminded me of myself in middle school, always hoping the ball wouldn't come my way. Miroslaw was eleven years old, with red cheeks and bright eyes, another child star born to the wrong epoch. Like many children, he stuck his tongue out of the side of his mouth as he drew. He beamed when Marko held his picture up.

It was a medieval scene. Rival armies faced each other beneath a stone tower. A man's head rested on a pike. A frightening figure stood beside him with an axe. The drawing gave off a melancholy feeling, part Edward Gorey, part Caravaggio (Figure 1). The children began to tell the story, suffused with laughs and shouts.

"The Muslims came to take the Serbs' land," Marko said.

"Murat," the girl, Katja, added, citing the name of the Turkish Sultan who led the Ottoman army onto Serb land.

Not wanting to be shown up—it was *his* drawing after all—Miroslaw quickly interjected the name of the leader of the Serbs, a noble called Lazar, a near saint in their eyes. The other youths repeated the two names, Murat and Lazar, and they sounded heavy with the repetition, shorthand for centuries of meaning to which I was not privy, to which I would never really be privy; this was not my story, as soccer was not my game. I don't know when, but at some point as they told the story, we started kicking the ball again. Mostly they kicked it to each other and let me listen and watch. Murat and Lazar, they said again. Murat and Lazar, who met in battle on Kosovo Polje. The words sailed with the ball across the grass.

There were other names, Vuk Brankovic, the traitor, and Milos, the hero. There are countless versions of this story, and they vary widely depending on who is doing the telling, an Albanian or a Serb. As these children told the story, it went like this:

Murat and his armies invaded Kosovo, which was the holiest land for the Serbs, the birthplace of the Serbian Orthodox religion. Monasteries and churches dotted the region. Many of them still stand today, though fewer after the 1998 war and the riots in 2004 that left much of the nation's treasures smoldering. Prince Lazar raised an army to defend the Serbian kingdom, but one of his noblemen, Vuk Brankovic, made a deal with the Turks.

"He was a traitor," Marko said with venom. "Without him maybe. . . ." But he didn't finish the sentence, the distant look on his face led me to believe he was imagining a Kosovo controlled by his people for six hundred years, a Kosovo where he was not in the minority, penned into fortified enclaves for his own safety, subject to the whims and rages of politicians and mobs. I wanted to ask him what he was going to say, but never got the chance. Eager Miroslaw continued with the story.

During the battle, a brave knight named Milos managed to trick his way behind Turkish lines. Pretending to offer himself in service to Murat, he knelt to kiss the sultan's hand. Instead, he stabbed Murat in the side, gravely wounding him.

"He did this because of Brankovic," Katja added. According to the kids, Brankovic betrayed Lazar, who was his father-in-law, by quitting the field at the height of the battle, thus allowing the Turks to penetrate the Serb lines, capture Lazar, and take control of Kosovo.

"They captured Lazar, see," Marko repeated, and pointed again to Miroslaw's drawing. It was Lazar whose head decorated the point of the pike. At his death, Lazar became a martyr.

"Tell him about the speech," Stefan said. Stefan had not spoken much since the story began. He seemed far more concerned with the mechanics of soccer than the details of history, but *the speech*, that was the piece that lit him up. He stopped the game again and held the ball under his foot.

"You do it," Marko said and Stefan did not need to be told twice.

"Before the battle," Stefan said, "Lazar spoke to his soldiers. He told them that he would fight for their God, and win the Kingdom of Heaven. Lazar said: It is better to die in battle than to live in shame."

Stefan was visibly moved as he spoke these words. The others nodded, and gazed at Lazar in the drawing, frozen in crayon, having lost his head, having lost his kingdom.

The way they told this story struck me in the same way they talked about the riots three months earlier that killed nineteen people. In mid-March, four Albanian children were playing by the banks of the fast moving river Ibar, near a Serbian village. The children entered the water, and three of them drowned. Immediately, rumors spread that the three children had been chased into the river by local Serb men with a dog. Speculation spread that that act was retaliation for the alleged gunning down of Serb children the previous summer by Albanian terrorists. Regardless, fury erupted in the Albanian community, with demonstrations throughout the country denouncing Serbian aggression. Those demonstrations quickly turned violent, and Serb homes and businesses became the targets of that violence. For the next three days, both Serb and Albanian mobs clashed, exchanging gunfire and tossing firebombs. More than 900 people were injured, 800 Serb houses, and 35 Orthodox churches were burned. Four thousand people lost their homes in three days.

These children had waited out the violence in their homes, nervously anticipating the arrival of an angry mob, but their homes were spared. The riots were over, and they had occurred for a simple reason, the children explained.

"The Albanians want to get rid of the Serbs so they can have Kosovo for themselves. That's what they've always wanted."

They told the story of the riots they survived with less outrage or animation than they told the story of the Battle of Kosovo that happened over six hundred years ago. It was as if the children had been there themselves in the summer of 1389, with their own heads on pikes, as if their own kingdom had been lost and the riots in 2004 were just aftershocks. Nothing was lost in March that had not already been lost on the medieval battlefield. In a sense they were right, as many see the Battle of Kosovo as the turning point when the Ottoman Empire took control of Kosovo, so that today Serbs are outsiders in their homeland. This is the magic that nationalism works on children. It was not an abstract historical wrong these Serb children felt. They still felt the hurt that began six hundred years earlier. They felt the hurt in their parents' humiliation, unable to find work in Kosovo. They felt it in the fear of Albanians, who surrounded them and penned them into the enclaves. They saw their current oppression as a result of their history, their ancient history. They made no mention of Serb discrimination against Albanians or of the campaign to cleanse Kosovo of the Albanians in 1998. It could well have been that they were kept somewhat ignorant of these events, as they would have been very small at the time. But Marko, at least, had lived in Pristina itself before the war. His family fled the anarchy and the reprisals against Serbs that swept the capital city after the NATO bombing allowed the Kosovo Liberation Army, the Albanian guerilla organization, back into Pristina to take control of the country and put an end to the ethnic cleansing. For the Serb children, the myths of their people from centuries ago were as real to them as the armored vehicles on the road by their school today.

What to make of these children in Kosovo who grew so moved by the telling of medieval history that they had to stop their game to make sure I understood their story, as if play would be impossible without this common narrative? What to make of these child soldiers in the Congo who roughhoused and laughed together only months after they had tried to kill each other? This was decidedly not how I imagined the children of war to be.

This project began when I was in college. It started with an insomniac night, when I watched reruns at three a.m. An ad for Save the Children came on, and I changed the channel. I did not want to sponsor a child, or even see those pictures the ad would inevitably show me. There was little escape, however. They were everywhere. The children were all the same: they were fleeing and hungry, all rags and bones and pleading, innocent eyes. They were from Ethiopia, Sudan, Liberia, Gaza, Afghanistan, Chechnya, Kosovo, Haiti, and Colombia. They were from the Congo and Sri Lanka. They were from Sierra Leone and Northern Ireland and from Israel. Where they came from hardly mattered to the story. The children were usually devoid of context. Seeing them I learned nothing about the conflict, the culture, or the child.

I had always thought of childhood as a kind of magical time, "a mythologized and privileged state," as the Oxford anthropologist Jo Boyden calls it, kept separate from the workplace, the world of adults, the hardships of the adult world. I held a conviction that in order to have a healthy childhood, the young must be sheltered from the struggles they would later come to know as grown-ups, that the most essential conditions for "normal" development were safety and stability. I also began to observe the widespread belief that children were not competent to face many of the harsh realities of life. There was a conception that children, by the simple fact that they are children, are "innocent in the ways of the world and incompetent in it."

Aside from the great pains that are taken to protect children from danger in the United States—look at the rubberizing of school playgrounds, the banning of tag, the flood of sanitizing gels in children's knapsacks, as if germs and skinned knees are no longer acceptable parts of play, and, most pernicious, the banning of books, and blocking of the Internet in an effort to protect the innocence of children—there is a near total denial that children are protagonists in their own lives. When a young person gets into trouble, blame is spread between the parents, the media, and any other cultural influence that is in vogue at the time: video games, loud music, fashion, MySpace. When a young person does something prodigious or remarkable—shows a selfless compassion by organizing a clothing drive for hurricane victims or, like twelve-year-old Ilana Wexler speaking at the Democratic National Convention, expresses a political opinion—it is seen as cute and somewhat unexpected, as if children were not usually aware of events in the world around them. However, it would take a supreme act of will for children in much of the world to be unaware of events around them.

Since World War II, children have become involved in wars in unprecedented ways. Jewish children were targeted by the Nazi death squads simply because they were children, and at the same time, Jewish youths fought with the partisans against Nazi occupation. Youths as young as twelve had to choose for themselves—pick up a gun and fight, or die in a gas chamber or ghetto. As David Rosen of Rutgers University notes, recruiters would often target orphans because their family ties had already been broken and they were less risk averse than children who still had family to lose. This practice is still in use today and is part of a well-developed doctrine of child-soldier use. Peter Singer of the Brookings Institute observed that there are established "best-practices" and global teaching pathways for the training and use of child soldiers, involving elaborate propaganda programs and ingeniously cruel desensitization regimes. These groups of armed children go on to terrorize the civilian population, turning expected social roles upside down. From Liberian street gangs and Palestinian suicide bombers to Afghan child laborers and underage Congolese concubines, children play central roles in modern conflicts.

Light weaponry, cheap to get and easy to use, has changed the way wars are fought, moving them from remote areas on the fringes of society to the center of villages and farms, the streets of cities. Worldwide, 2 million children died as a result of armed conflict in the 1990s, more than 20 million children were displaced, uprooted from their homes by violence and forced to flee. More than 6 million were disabled or wounded, and an estimated 300,000 were recruited into military or paramilitary forces as soldiers, porters, cooks, minesweepers, sentries, spies, or sex-slaves. Children, especially adolescents, have become more central to the way wars are fought, as targets of violence and as combatants. Their involvement in modern wars cannot be classified as passive.

Children are, rightly, of great concern to any society, but because of that, they become its most often used (and misused) rhetorical tools, its obsession. The word "children" is invoked to support all kinds of political agendas, depending on the need of the individual or group invoking them.

In the madness of modern warfare, there is a method to the exploitation of children. In the conflict between Palestinians and Israelis, pictures of dead children are displayed as a call to arms, to continue the political struggle in their name. I still marvel at funerals in which angry demonstrators carry large placards bearing the photo of the deceased child, just days after the child died. How did they get the photo so fast? How did they have it enlarged and printed and distributed so quickly? Someone must have gone to the relatives immediately, looking for a good martyr photograph of the child, or the relatives must have thought to present it right away.

When Israeli shells killed seven members of a Palestinian family on a beach in Gaza in June 2006, Palestinian leaders did not hesitate to turn the one surviving child, seven-year-old Huda Ghalia, into a potent symbol of the conflict with Israel. Mahmoud Abbas and Prime Minister Ismail Haniya symbolically adopted the girl within hours of her parents' deaths. While the funeral was underway, Hamas renewed its rocket attacks on Israel in retaliation.

In nationalist struggles from Rwanda to the former Yugoslavia, politicians have called on the people to rise up against their enemies in order to protect their children's future, their children's rights. In the prelude to the genocide in Rwanda, rumors were spread that Tutsis had attacked children at school and were working together to prevent the Hutus from having a future in Rwanda. The next step for any Hutu interested in protecting his children's future was to attack and eliminate the Tutsi completely.

In Bosnia, during the siege of Sarajevo, people risked darting into sniper zones to drag injured children to safety. Also during the siege, young girls were targeted for gang rape and brutal torture in order to demoralize the entire society. When a shell was fired into the yard of a Sarajevo kindergarten, killing several small children, a friend told me with horror that the shell casing was engraved with the words: "A hot kiss from us to you." How she heard this, I do not know. It might be apocryphal, but that she used it to illustrate the worst horrors of the siege, when there were plenty of verifiable horrors to describe, showed just how terribly war crimes against children shake people.

In the 2006 clashes between Israel and Lebanon, one medical organization stated with alarm that a "disproportionately high" number of children were endangered by the conflict, at risk not only to the falling bombs, which killed three hundred children and injured thousands more, but due to serious health problems that would develop even after the fighting ceased. Without a doubt, during wars, children are victims.

But what of the myriad children who care for their cousins, brothers, sisters, and uncles in impossible conditions? In my time doing research for this book, through innumerable soccer games and melted crayons and slow walks with young people, I had the privilege to meet children around the world who have survived and continue to survive unspeakable horrors and chronic deprivations. Each one of them has a unique genius for survival, physical and psychological, sometimes with the help of adults, often on their own, and many of them flourish: they manage to eat, to play and laugh, to help others, to find support when they need it, to make challenging decisions when they have to. It would be both presumptuous and meaningless to say that they did not have a "childhood" because they did not grow up in the rather unique safety and stability common to Western notions of child development.

Neil Postman asserts in his book *The Disappearance of Childhood* that the modern world is "halfway toward forgetting that children need childhood," by which he means that the modern world is forgetting that children need to be kept separate from the adult complexities of life in order to be healthy and happy. Though his was largely a critique of media culture, that same notion has been exported to humanitarian crises as well, taking the basic assumption that children occupy a privileged space in society as gospel truth. But this privileged space has never been a given in much of the world, and the notion of childhood has never been a fixed state.

The United Nations, in the Convention on the Rights of the Child, states, "a child means every human being below the age of eighteen." As a legal instrument, this is sufficient, but it sheds no light on the various cultural contexts that define the relationships between the world of children and adults around the globe.

For the Gussi tribe of Kenya, the rite of passage into adulthood occurs at age eight, while the Jewish bar mitzvah occurs at age thirteen. In some societies childhood ends for girls with puberty and marriage, or is defined by whether or not a boy is in school or working. In medieval Europe, childhood ended at age seven, when children could master spoken language. The Catholic Church still recognizes age seven as the age at which children can understand concepts of right and wrong, can begin to use reason. As St. Thomas Aquinas wrote: "Age of body does not determine age of soul. Even in childhood man can attain spiritual maturity."

Childhood is not a concept set in stone. The idea of childhood, as most of us understand it, has existed for less than two hundred years. In the Middle Ages in Europe, no distinction was made between children and adults. They wore the same sort of clothes and once they could care for themselves, as far as we know, they did. They worked and talked like adults. They did everything adults did. Childhood was given no special place in society, and children were not particularly valued. Their death rates were so high that few got very sentimental about childhood. In a seventeenth century French manuscript quoted by Philippe Ariès, in his *Centuries of Childhood*, a neighbor consoles a young woman who has just given birth to her fifth child by saying: "Before they are old enough to bother you, you will have lost half of them, perhaps all of them."

In the Middle Ages, there was no great effort to protect children from the adult world, as there was no concept that they were not already part of that world. Childhood sexuality was not taboo. It was, in fact, a source of great amusement to all levels of society, as Ariès notes with a particularly graphic account from the life of young Louis XIII. Looking at paintings of peasant life in Holland by Pieter Brueghel, one can see children fully engaged with the adult world, in feasts and drinking and even in the march with Christ to Calvary.

The idea of children inhabiting a privileged sphere of society from adults only emerged in the seventeenth century, and even then it took root only in the middle and upper classes. As literacy attained value in Europe, children needed to acquire the necessary skills to become literate, which led to greater interest in schools than ever before. Educational standards became necessary and schools became the place to instill the standards. Schools were separate from the world of adults, and pupils were expected to study and learn, rather than to work. They were to learn to become adults, to gain access to knowledge that would make them able to participate and succeed in the adult world, which had become a world filled with information thanks to the printing press. School was a place for children, the world beyond a place for adults.

Of course, in order to send one's children to school, before free public education was commonplace, one had to be able to afford it. One had to remove potential contributors to the household economy from the equation while taking on an extra expense. Among the lower classes in Europe, this did not happen. Childhood took much longer to trickle down to them, and the young remained fully immersed in the world of adults, with no distinction, as it had been for centuries. The idea of childhood was a luxury for the wealthy and the safe. During the industrial revolution it was the elites who protected the notion of childhood, even as the children of lower classes worked in deplorable conditions for obscene hours and low wages. The idea of childhood was nearly lost in Europe. The elites led the drive for policy changes that protected children, and the elites consumed the goods—clothing, books, games—created for children. It was a long struggle to create a culture in which the idea of childhood could be taken for granted by all levels of society.

Today, these conditions are not universal. Economic, cultural, and political factors clash with the imported idea of childhood on a regular basis. Economically, it is clear why, in very poor countries, children would be compelled to work—tending the flocks or farming or making carpets or begging to help support the family.

Even with access to schooling, in areas of extreme poverty, the Western ideal of childhood is not secure. In Senegal, it is a common practice for young boys to be sent away to religious schools, called *daara*, where they are sent out during the day to beg for their teacher. If they come back empty-handed, they are beaten. As one boy, Ibrahim Sow, a sixteen-year-old in Senegal, told the IRIN news service:

At the *daara* I used to get up at six a.m. and go out to beg for my breakfast as there was no food there. At nine o'clock I'd return to learn the Koran until one p.m., when I'd go back out to beg for my midday meal. I'd return to the *daara* at three p.m. and stay in class until five. It was at five p.m. every day that I had to turn over all the cash begged that day. There was no amount set, but when we came back empty-handed we were beaten. They only let us buy food if we brought back a lot of money. Otherwise we didn't eat. The toughest times were when the *marabout* teacher was away, because then the oldest *talibes* were left in charge. There were only five or six of them, but they didn't treat us well. That's why I ran away from the *daara*.

After running off, I lived on the streets for two or three months. Sleeping rough wasn't easy, but I was never scared. I used to sleep under trucks or buses at bus stations. I'd chase away anyone who was already there to get some space. I used to beg and steal, and I was never caught, except for once . . .

me and a friend stole a cell phone, and the owner saw us. He caught us, and heated up a fork and a knife, and then applied them to our skin. I still have burn marks on the stomach, chest, left arm, and bottom. One day I sniffed paint solvent with a gang of friends. We put the stuff on a rag, and mixed it with fresh mint to hide the smell. It had no effect on me, so I threw away the rag, and never tried again. I used to make love to younger boys, but nobody ever did it to me. I had seen a friend of mine who's leader of a gang of street kids do it, so I tried. I dropped all that when I came to live in the shelter.

Cultural differences also clash with the Western idea of childhood, given the different ages and circumstances that define a child's relationship to the adult world through coming of age rituals and so forth. But politics too can dash the fragile and hard-earned status of the young against the rocks.

Consider little Johnny and Luther Htoo:

In January of 2000, a group of Burmese rebel fighters from the insurgency group God's Army took over a hospital in downtown Ratchaburi, Thailand, holding hundreds of hostages, sending Burma's civil war spilling over Thailand's borders. The Karen are a minority ethnic group in Burma, primarily Christian. They have a large and sympathetic population along the Thai-Burma border.

"We want to tell the world how Karen and Burmese refugees live during the fighting. We will not hurt any of the hostages, we will take good care of them," said one of the gunmen to Reuters Press Agency. The rebels were fighting as part of a long war against the military junta in Burma (Myanmar, as the ruling junta renamed it in 1989—the names matter, the names tell whose side you are on, where your loyalties lie). The junta has one of the worst human rights records in the world: they routinely jail dissidents, censor the media, target and displace civilians in order to cut the rebels off from support, and, according to reports from several human rights groups, use rape as a tactic of war. They justify their actions by claiming the rebels want to break Burma apart along ethnic lines, though the various ethnic armies have long disavowed that goal.

Living in New York City, it is easy for me to condemn violence and espouse pacifism, to urge nonviolent resistance, but I can understand how, for someone living under the conditions created by this regime, fighting against it could be a valid choice. A reasonable person could rationally choose to fight this regime. The question comes down to how much credit we give young people for rationality.

God's Army, which was based in the jungles near the Burmese border with Thailand, was led by two twelve-year-old brothers from the Karen ethnic group, Johnny and Luther Htoo: black-tongued, cigar-smoking soldiers whose followers claimed they possessed magical powers to dodge bullets and step on land mines without setting them off. Their legend began when their village was attacked. By some accounts, they rallied the villagers around them and fought back, killing many soldiers in the process and escaping unharmed through a heavily mined area. They were nine years old at the time.

The God's Army group that took hostages in Thailand three years later said they wanted attention rather than power. They wanted doctors and nurses to fly to the border regions and aid their injured fighters. They demanded that Thailand stop aiding the Burmese government's attacks on rebel bases. The hostage-takers freed fifty of their prisoners after the Thai army chief ordered his troops to end the shelling of the guerillas' base on the Burma border. The only casualty occurred when a stray bullet injured a teacher at a nearby school, according to a local radio station. The next day, the Thai army raided the hospital, killing ten of the gunmen. The *Sydney Morning Herald* reported, "Some of the bodies [of the gunmen] were reported to be small and childlike."

I think Johnny and Luther would have identified with the rest of the quotation attributed to St. Thomas': "Age of body does not determine age of soul . . . many children, through the strength of the Holy Spirit they have received, have bravely fought for Christ even to the shedding of their blood."

In 2001 Johnny and Luther turned themselves in along with seventeen of their fighters. Among them was a bodyguard named Rambo, who many believe was one of the real commanders of the militia, using the boys as cover, training and guiding them while cultivating their public image. Many on both sides of the border, however, still maintained their faith in the two boys as commanders and saviors.

"I want to live as a family with my parents. I want to study," Luther told the gathered press when he and his brother emerged from the jungles to end their days as soldiers. Their militia was in shambles, starving and under constant attack by the army.

Entering Thailand carrying small backpacks, looking frail and tired, they were received by the Thai prime minister, who promptly patted them on the head, a social symbol in Thailand that immediately reduced their status to that of schoolchildren. One would not pat a god on the head.

The boys explained that they had no magical powers. They found their mother in a refugee camp in Thailand. In 2004 sixteen-year-old Luther Htoo married a nineteen-year-old woman and they had a child together in the camp. Luther says he wants a job so he can support his young family.

Johnny says he still thinks about the oppression of his people. In 2004 he told a reporter for the Bangkok-based *Nation* newspaper, "If I could, I would exchange a comfortable life [in the camp] and die for the peace of the Karen nation."

He may have gotten in his chance. It is possible that, at eighteen years old, he left the safety of the refugee camp to return to fighting. A junta-backed news service in Burma reported on July 26, 2006, that Johnny Htoo and several other followers of God's Army "turned themselves in" to the military junta. The truth of this claim is a mystery, as the media is tightly controlled, but indications suggest that Johnny was no longer in the refugee camp where he had been living. Whether he turned himself in (which is doubtful given the junta's reputation for torturing prisoners), or was captured is unknown, but as a young adult he had made the choice to return to the war he had left as a child while his brother stayed behind.

Johnny and Luther Htoo stir up questions for me. What makes them so different from the kids I've seen in pictures on the news from refugee camps, the victims of war? What makes them so different from American children? What makes them so different from adults? They were physically capable of anything the grown-up soldiers could do. Johnny and Luther clearly had political views, or at least political impulses; they made moral choices, perhaps not always good ones, but regardless, they made commitments to their people and their cause, and they are growing up making choices based on value systems formed when they were younger. If children can make such commitments, what does that say about their childhoods? Were Johnny and Luther robbed of their childhood by the adults who drafted them into the fighting or turned them into messianic symbols? Would it have been better for them to grow up complacent, live under the regime that oppressed them, and remain ignorant of their people's struggle until they turned eighteen?

One could say their less than ideal conditions forced Johnny and Luther to become soldiers and killers, and this brings up an essential question for me:

When were safety and stability the norm for young people?

Why is it essential to childhood development that safety and stability reign? Doesn't human history show us that war, violence, loss, and upheaval are far more common than peace and prosperity? It is estimated that since 3600 B.C. there have been 14,000 wars, and at least 160 since the end of World War II. Looked at in this light, children who grow up without knowing war would seem to be the exception in history.

Though forged in violence, Johnny and Luther Htoo have distinct personalities with different goals and values. Which one of them is "healthy"? Luther, who wants the peace of a family life or Johnny, who carries on the fight to liberate his people? Even barring political violence, there are little earthquakes that much of the world's young must withstand: hunger, poverty, domestic violence, divorce, disease, crime, urban blight, rural isolation, and so on. Not a sunny picture of the world, but a picture that is fairly common. Do all children who grow up in tumultuous situations have developmental problems? Do these realities rob them of childhood?

It is true that in early childhood, the ability to care for oneself and solve complex problems is not developed, and young children need the support of parents, caretakers, or older children to provide for their needs and shield them from hardship as much as possible. But extending this view into adolescence— middle and late childhood as some researchers call it—might not be helpful in understanding the way these adolescents experience and adapt to political violence, displacement, and loss and may actually *harm* their ability to cope with these stresses by encouraging passivity and disengagement. For this reason, I decided to focus on adolescents in my travels, roughly children eleven to eighteen years old.

Anna Freud, perhaps the most influential figure in the field of child psychoanalysis, once said: "Let us try to learn from children all they have to tell us, and let us sort out only later how their ideas fit in with our own." It is with that statement in mind that I came to this work, hoping to listen and learn who these children of war really are. And there is no better way to begin to learn than through play.

Sitting on the sidelines after an hour or two of soccer skir-mishing with the child soldiers in Bukavu, I began to speak with Paul, the thoughtful boy who had helped me up from the mud and guided me through the rules of the game.

Paul was about four feet tall, but said he was fifteen years old. It is possible. Due to malnutrition, many children in impoverished and war-torn regions suffer from stunted growth, but another explanation seems likely to me in retrospect. He was not fifteen years old. He was eleven or twelve years old. Lying to me was part of a well-learned strategy for survival.

Under international law, the minimum age for participation in the armed forces is fifteen. Recruiting children below the age of fifteen is considered a war crime. I believe Paul was instructed that, when asked his age, he should always answer "Fifteen." He followed these orders because to disobey an order is very dangerous, especially for a child subject to the will of the older commanders. Discipline for insubordinate soldiers is always fierce when there is a war on.

In Colombia's rebel movement, FARC, for example, insubordinate child soldiers are tied to a pole or a tree by a length of nylon cord. They are forbidden to speak or be spoken to. According to Human Rights Watch, children can be left tied up like that for weeks or a month, thirsty and hungry and alone. At times, their peers are called on to execute them. They are ordered to beat the restrained child to death with stones or sticks, teaching those who live a hands-on lesson about insubordination. The luckier ones are summarily shot by their commanders.

But Paul might have maintained he was fifteen for reasons other than fear. In the army, he was clothed and fed. In a land where there is little employment, where the availability of regular schooling is not always guaranteed, having food, shelter, and activities of a kind can seem like a great opportunity for a young person. To stay in the army, you had to be fifteen, so Paul became fifteen in the hopes of having a better life.

There are not many birth records in the Congo. He may not have known his age at all, and being told he was fifteen was all he needed to believe it. By being fifteen he was part of a group, was given a purpose and accepted, and protected himself from anyone checking to make sure he was following orders. Paul could not be sure what my intentions were in interviewing him, despite what I explained to him, and he was being very cautious. In fact, all of the littlest former child soldiers I met told me they were fifteen. Every single one.

Paul was living in the center waiting for his family to be located to begin the process of reintegrating him into civilian life. It was uncertain if his family would be found, the caretaker told me, and if they would want him back. Before arriving at the center, he was fighting for the Mayi Mayi in the eastern Democratic Republic of Congo.

The Mayi Mayi is a name given to a number of militias and local defense forces fighting to push foreign invaders out of the Congo or even out of whatever area a particular militia calls home, though they were often no more than violent groups of local thugs taking what they wanted from the people with the barrel of a gun. Mayi comes from the Swahili word "*magi*," which means "water." Local medicine men would encourage youths to join the post-colonial militias, blessing them so that any bullets fired at them would turn to water. Their chant evolved to Mayi Mayi, and the popularity of this belief spread. The power to turn bullets to water proved not to be within the various commanders' and warlords' power, but the name stuck. The Mayi Mayi are notoriously fierce, and their command structure is loose. They are feared and often blamed for a lot of the massacres in the eastern part of the Congo. Most factions are organized along ethnic lines, and their violence is directed toward other ethnic groups whom they see as invaders, specifically the Tutsi from Rwanda and their ethnic counterparts in the Congo.

When I met Paul in the winter of 2002, the war was still tearing the eastern Congo apart, and the rebel group RCD-Goma controlled much of the region, with the backing of neighboring Rwanda. Two Rwandan soldiers stood about 100 yards down the road from where Paul was staying, watching the children come and go, ready to take some of them back into the army. They passed the hot days using their pickup truck for shade. The threat of a kidnapping raid loomed over the center.

I told Paul that I would change his name in anything I wrote down so that he could not get in trouble for what he said to me. We could pick his pretend name together, and he could say anything he wanted without being afraid.

He looked at me with wide brown eyes and an expression of confidence on his face, a smile, because I did not seem to understand anything and he would help clear things up for me.

"It's okay. I am a soldier," he said. "I can't be afraid." I think he would be disappointed to learn that I changed his name anyway, as I did for all the children in the interest of their own safety.

Paul was thoughtful and eager to help the other children.

"There were twenty children fighting with me. They wanted to escape from the army too, but only five of us decided to try.

The others said, 'If you arrive safe, let us know, so we can escape too.' I have not been able to send them the message yet." It was his "yet" that struck me. He had plans to help his peers, which he intended to carry out.

Paul had been kidnapped into the army; he was not a volunteer, he said. One night, the *interhamwe* entered his village. The *interhamwe*, roughly meaning "those who attack together," were the militia in Rwanda primarily responsible for the killing of nearly 800,000 people in the three months of genocidal fever that swept the country in April 1994. They were police officers and soldiers, farmers and businessmen, teachers and students, nuns and priests and nurses, driven mad by ethnic hatred and manipulated by their leaders. Most of the murders of Tutsis and moderate Hutus who refused to go along with the killing were committed with machetes and garden hoes. After the war, the *interhamwe* fled Rwanda into the jungles of the Congo where, still manipulated by the former leaders of the Hutu Power government, they continued to stage attacks against Rwanda, continued to avoid punishment for their crimes, and continued to profit from the resource rich land of the eastern Congo for over ten years.

In Paul's village, everyone was sleeping as the *interhamwe* crept in through the jungle. They burst upon the village, ripping the night open with machine-gun fire. Women and children were roused from their beds. The *interhamwe* looted food, blankets, and supplies. The village, already poor, lost everything.

As the village burned and families ran for shelter in the bush, the soldiers discovered Paul. They grabbed him and demanded that he help them carry their spoils. He picked up the food, radios, and cartons of cigarettes that they handed him and followed them into the jungle, away from his village. He never saw his mother or father to say good-bye.

On his trip through the jungle, he complained that he was very far from home. The soldiers looked at him coldly.

"You are one of us now, a soldier," they said. "You cannot go home. Even if you wanted to, they would think you were with us and accuse you of stealing. You are no longer a civilian."

During his time in the army, Paul says, he fought for the Mayi Mayi faction led by Colonel Padiri, whose nom de guerre means priest. I could imagine the sermons coming from this "priest" as Paul spoke.

"We were fighting to liberate the country from Rwanda's army." Though the Mayi Mayi and the *interhamwe* are separate entities, they both have the same goal: the expulsion and downfall of the Rwandan presence in the Congo. The *interhamwe* must have handed Paul over to the Mayi Mayi so they wouldn't have to feed him.

He cannot count the number of times he was sent to the front lines. "I went to Kambegeti, Bulambika, Nyakakala four times, Mubuezu five times, other places. We would fight and run, not take villages. We would shoot at the Tutsi [Rwandan soldiers] and run away."

As Jason Stearns, a senior analyst with the International Crisis Group, told a reporter, "Padiri's Mayi Mayi were guilty of widespread rape and abuse [against civilians]."

Paul had clearly adopted the party line about the Rwandans being foreign occupiers, about the Tutsi ethnic group as the enemy, but when I asked him about the cause of the war, he told me it was the fault of the *interhamwe*, his abductors, because Rwanda fought them here. He seemed to understand the necessity of one group fighting the other, or at least the inevitability of it. He did not really care about the who's and why's of the fighting. He wanted only one thing: to go to school.

"My parents could pay my school fees. They could afford

it and I could go back to school. I only went through primary school."

Paul wanted to be a student, to return home to his parents, put on the white shirt and blue shorts that schoolboys wear, and study. Many of the child soldiers I met, like Paul, wanted only to study. Two themes dominated the drawings child soldiers made for me: violence and school.

Their drawings of violence were often detailed and large, with big guns in the pictures, the make of the guns, the texture of the camouflage thoroughly depicted. The pictures of school tended to show less attention to detail (many of the kids had not been to school for a long time), but the children usually told me that they themselves were in the drawings of school, not in the drawings of violence (Figures 2, 3).

School held a strong allure for most of the children I met; they longed for the classroom the way children in the West tend to long for the playground. There are a variety of reasons for this, but in most cases I imagine it had to do with two major factors: a break from the harsher world outside of school and the belief that through learning they can improve their situation in life. For most children affected by war (indeed for many poorer children not affected by war), life consists of a hard regime of work. One tends the flocks or works to earn money for the family or helps with the cooking and cleaning and carrying water and wood. Out of school, one is exposed to all manner of perils, the same perils that face grown-ups, and one must usually face these perils alone. In school, however, a child can sit and play; can spend time with other children and not have to do any of the labor that comes outside the schoolhouse walls. One's only job in school is to learn. Children in school socialize; they feel like part of a community, a feeling that is all too scarce during times of war. Socializing in school is an immense privilege for children who have spent most of their days as soldiers or beggars or touts for busses and brothels or as prostitutes. School provides a level of stability that most young people long for and that most young people experiencing war cannot find anywhere else.

In addition to the benefits of being in school, children want the benefits they believe school will provide. They want to learn skills that will create opportunities; they want knowledge, which they believe will give them choices in life. For children of war, school becomes a pathway out of the present suffering and into the future, a future filled with stability, safety, and prosperity. Unfortunately, school is not a possibility for most children in regions affected by armed conflict. In the Democratic Republic of Congo only about 35 percent of school age children attend school and that percentage is no doubt lower in the war-ravaged east of the country. Paul and other former child soldiers want the future that schooling can give as a way to avoid the future that they know soldiering *will* give them. From Paul's point of view, without an education he will have no hope for a future that is any better than his past.

Paul taught me a great deal about children in situations where violence and hardship are the norm. He showed me that the young have the capacity not only to survive but to act with altruism. In trying to help others escape from the Mayi Mayi, he was exercising a moral will that fails many adults in dangerous situations. He had learned the political rhetoric of his militia and seemingly understood it, yet he could be critical of it, framing it first from the point of view of his commanders, and then, later in our talk, considering the Rwandan position, that they needed to fight the *interhamwe*.

I thought about our soccer game again. Paul's warm consideration for the other players struck me profoundly. I wish I had asked him more about his feelings for the other kids in the center. I can only make assumptions from what I saw when we played, but his sense of the situation of others, whether they were the enemy army or other players fallen in the mud during a game, suggested a possibility I had never before considered: through the crucible of violence and hardship, some children can develop deep moral sensibilities and can flourish as individuals.

"I regret my time in the army," he said when I asked him what he would tell the leaders who caused the war if he had the chance. "I would like to say to other children not to join the army. To those who cause war I have nothing to say because I am too young. If I had power. . . ."

His voice trailed off here, and he looked at the floor. Paul seemed to recognize the sidelining of children's views and experiences. "I am too young," he said, but I wondered how that could be possible. Since he had fought in the war, I did not believe he was too young to have an opinion. He had been expressing complex ideas all morning.

He screwed his forehead in thought, his eyes focused past his shoes, seemingly through the floor at the mud just below the boards. I waited for him to continue. It had started to rain outside, and the drops played the tin roof of the center like a xylophone. I wanted to know what Paul would do if he had power; I wanted him to finish his sentence. Would he get revenge? Would he outlaw war? I thought these answers would tell me a lot about his "psychology." He looked up and told me his answer in a level voice.

"Everyone is killing people, dying for nothing."

Paul understood the reality of the war in the Congo. He understood it through all the propaganda pumped into him by the army. He didn't know the nuances of the political scene, the names of the players, or the interests profiting from it, but he understood an essential factor of the war: the reality for most Congolese civilians was that neither the victims nor the killers were fighting for any reason anymore.

After five years and an estimated four million people dead, the war in the Congo was declared over in the spring of 2003, though sporadic fighting continued in the east and the threat of renewed full-scale war looms. The violence of the war has its own momentum, not so easy to stop with treaties. Even in 2002, when we met, when the end of the war did not seem anywhere in sight, Paul did not want to discuss it much. There was not much point in discussing it, he told me. His worries when we spoke were about school and his future.

Paul took an active interest in what would happen to him next. He expressed hope for himself if he could go to school, hope that he could have a good life, as a mechanic, he suggested, if he could get out of the center and go to study.

"There is always work for a mechanic here," he said. "The roads are bad, cars are always broken." He smiled because he had seen the car in which I pulled up, a busted up taxi that my translator (who was a cab driver too) borrowed to take me around. When I mentioned our car, he laughed and suggested to the translator that he let him fix it, even though he had not been trained as a mechanic yet.

I didn't really know how to think of him at first, this little boy who was quick to laugh and smile, who had fought with one of the most brutal militias in the world, with some of the worst killers of the twentieth century, who called himself a soldier and denied that he could be afraid, and who desperately wanted to leave the life of war and go to school.

He had a lot of self-confidence, which I was inclined to call a defense mechanism against all the stresses he had experienced.

As defense mechanisms go, it seemed a pretty reasonable one. He had learned not only that he could not rely on adults to have his best interests at heart—adults had abducted him, after all—he had learned that he could have power over adults when he was a soldier. He could take control of his own situation, for good or ill, and behave like an adult himself. Considering his regard for others at the time we talked, I hope he will continue to nurture the impulse towards kindness. It could easily go the other way. He had been trained to kill and told it was okay to do so. I have no way of knowing what happened to Paul—fighting in Bukavu has displaced many residents since we met. He could have been compelled to rejoin the militia, or targeted by another army. He could have stopped being thoughtful of other children, started getting in fights. I like to think he got what he wanted, the chance to go to school.

"I don't like it here, in the center," he said over and over. "I want to leave. I don't have anything to say to you except that I want to leave here and to study." I imagine he did not like the confines of the compound, nor the boredom and uncertainty of waiting for parents or other relatives to be found.

Paul was vulnerable for recruitment because he was physically developed enough to carry a gun and because he did not have the same rights or protections as an adult (though adults are also forced to join armies). He was recruited *because* he was a child, not *in spite of* being a child. When the soldiers told him he would be accused of helping them steal, to whom could he appeal? Adolescents are often mistrusted. Paul sensed his position was weak and the soldiers were in a position of strength, so he made what I see as a survival calculation: stick with the strength, go with the soldiers. Otherwise, become a victim, either of the soldiers or of your own community who will suspect you.

He could be filled up with political propaganda because he respected authority and understood what was expected of him. After escaping the army on his own, failed by adults who could neither protect him from the military nor get him out of it, he found himself in a center waiting for other adults to come to his aid. But I believe Paul is quite capable of helping himself, if his altruistic impulses can be encouraged and he can have access to the resources he needs to get an education. He has not given up on the world or on the adults around him. He wants their help and is waiting for it, against the odds. He wants school and parental guidance, not the adulthood that was forced upon him by soldiering. I respect him not for the choices that were made for him—joining the army, waiting for help in the demobilization center, but for the choices he has made: kindness toward others, hope for the future, a desire to learn.

The lesson I took from Paul was this: In wars, when the world of grown-ups fails them, some kids can create their own conditions for survival, can help others to survive, can show amazing courage and strength, can carry the burdens placed on them for quite a while. They are capable of this and deserve, in fact need, respect and encouragement for these capacities. But it is up to adults, who are far more culpable in the political realities of the world, in creating the environment from which children learn to act, not to allow children to carry these burdens for long. They do not all hold up under the strain like Paul.

*This project is the result of research missions in East Af-*rica, Thailand, and the Balkans. It is by no means a complete picture of the impact of war on children: I am no expert and my regional scope is limited. I worked with a translator most of the time, and in some cases this translator had his or her own agenda. I tried to render the children's words as faithfully as possible and made every effort to work with translators who had the skill and the ability to render the sense of the children's words without editing them. This was not always possible, and there were times I might have missed what a child was actually saying or actually meant. I can only ask that the reader trust what I convey, as I had to trust what I heard and saw.

I have changed the names of the children and other people involved in these conversations in the interest of their safety. I have also changed a few details to further mask identities where appropriate, and the order of a few events for narrative clarity, but nothing of substance has been altered in their comments or stories. The conversations I recount are told much as they occurred, though, by necessity, many of the meandering discussions off-topic have been edited out.

Follow-up with the same children over time was possible only in a few cases, and I realize it is nearly impossible to predict how a child will grow up based on a few interactions, conversations, and soccer games.

I am not trying to come up with a general theory of how young people experience and cope with war. Anything that is true for one child in one conflict may not be true for another. Differences in culture, political structure, age, gender, and the social status of the child affect responses to stressful and dangerous situations to a great degree. I hope only to document a few young lives that have been touched by war, to pay my respects to their survival and to applaud the often startling intelligence and resourcefulness of young people who do get through war and can flourish afterwards. I will try to highlight the factors I noticed that might make a young person more resilient in war, but these are by no means "scientific" observations. I hope to dispel the notion that young people are passive victims, vehicles for suffering, as they are presented in most news reports. Children are protagonists in wars, from Angola to Iraq, with their own needs and desires, and they cannot be ignored.

TWO

"Then He Lined Us Up"

Children Fleeing

I n his *Decline and Fall of the Roman Empire*, the eighteenth-
century British historian Edward Gibbon describes the scene
of Romans fleeing the city of Nisibis in A.D. 363 after it was
handed over to the Persians: "The highways were crowded with
a trembling multitude: the distinctions of rank, and sex, and
age, were lost in the general calamity. Every one strove to bear
away some fragment from the wreck of his fortunes. . . ."

Gibbon could have been describing a photograph from the
1994 genocide in Rwanda or the 1998 campaigns of ethnic
cleansing in Kosovo or the crisis in Darfur, Sudan. He could
have been describing any number of forced mass migrations
that have occurred all over the world in the last ten years, even
the last five. The picture has not changed much since the fourth
century.

Gibbon could have been describing the drawing that Keto,
a fifteen-year-old Congolese orphan, made for me under a
thatched roof in Lugufu Refugee Camp in Tanzania, where he

had lived for three years. As he drew, others came over to look at his drawing and he shooed them away so he could concentrate. I watched him gaze up at the roof while he drew, playing out the picture in his mind.

He labeled his picture, "The War in the Congo," and in it he depicts his escape from the war zone.(Figure 4). At the top of the page, in the mountains, a road begins. This road crisscrosses the page all the way to the bottom, taking a circuitous route past a helicopter that is dropping bombs on the fleeing civilians. The road opens out at the end of the page, wide and full of possibilities. Keto has made the road to that point as long as it could be on a piece of drawing paper, zigzagging from one side to the other. People bearing loads on their heads are rushing down towards a flag from which a boat is leaving, also packed full of people. Along the road, there is a dead stick figure, his head X-ed out in blue. A blue X also crosses his knee at a point where it bends off at a sharp angle. Next to the figure is the dropped load he was carrying; I wonder if Keto is depicting the actual wounds of a man he saw.

"She's died by the side of the road," Keto told me. "She was killed by the Mayi Mayi." The figure had no gender markings—I assumed it was a man—nor any distinguishing features of any kind save the blue X's, yet Keto seemed to be thinking of someone specific. In his mind, the wounds were the most distinguishing features of this woman, all he chose to depict, perhaps all that he remembered.

He seemed frustrated at our discussion of his drawing. I'd only known him for about an hour. Keto was the first boy I met in the refugee camp, the first Congolese child I was meeting in my life, the first person I'd interviewed about his experiences of war. I was nervous and did not want to frustrate him. I wanted him to like me. My mind raced. He was very quiet. He said something quiet to the translator. I worried that he might be traumatized from his experiences, and I did not want to open up wounds in his mind and then leave him to suffer the consequences of them while I got back in the white UN jeep and drove away. I decided to change the subject, to talk to him about soccer, because he had also drawn a picture of a soccer ball.

"I like football," he said, "though there are not enough balls here in the camp."

I began to ask another question, neglecting whatever connection may have formed had I allowed soccer to take center stage. In my eagerness in this interview, I wanted to ask some *revealing* question, questions that would get to the *core* of Keto's *being*. I had not yet realized that soccer could be the key, that play could reveal the secrets of Keto that words would not. His answer about the number of soccer balls contained a universe of information about how he felt, what he wanted, what he hoped to get from me. I plowed on, oblivious.

"Keto, can you tell me—"

My translator stopped me mid-sentence and paused for a moment. He turned to me. Keto was not going to let me get by that easy, not going to let me miss the connection his answer about soccer balls demanded.

"Before you go on, Keto would like to ask you a question, if it is all right."

"Of course," I said. "He can say or ask anything he likes." I smiled to show without words that I was very happy to answer his questions. The interview still felt more formal than I had wanted my interviews to be. It would take a bit of practice, letting a conversation flow between a child, a translator, and me.

Keto sat up straight and looked me right in the eyes to ask his question.

"How will talking to you about the war help me to get shoes

or more food or a blanket?" *Or more soccer balls*, his question seemed to say. Perhaps he was too polite to throw that in as a dig against my obliviousness. I had much to learn.

I was not sure how to answer. This was a question I had been thinking about since I thought of this project, since I arrived in Africa for the first time a few days earlier. It was a question that would haunt me for the next three years as I returned to Africa to meet other children, as I met children who had become illegal migrants in Thailand to escape the junta in Myanmar, as I met orphans struggling to grow up and build their lives in the stunted economy and traumatized villages of post-war Kosovo. The implications of this question harass me as I write this now.

Anyone who does "field research" (I hate that term; implying pith helmets, Stanley and Livingstone) in communities that are less fortunate than one's own—whether it be documentary work, social science research, humanitarian assessment, anthropological study, or journalism—has to deal with the moral and psychological tensions, as Robert Coles calls them, that this kind of work creates.

How do you arrive as an outsider among people who are struggling to survive, observe and interview them, take their lives as "material," and leave? If you are successful, it will be in large part because of the quality of the material (the content of the lives researched) that you have gathered. Your career will advance or your reputation will be made. Yet what of the people you have observed or, to put it another way, *exploited*? How does telling their story help any of these children, and how can I sleep at night having taken their stories and left them in war zones, still hungry, still at risk?

There are no easy answers to these questions, and few resolutions to the tensions. I believe there is value in being heard, in sharing your experiences with others. There was a value for the youths and adolescents I met while doing this work: I communicated their concerns to those who might make changes, I helped to validate their thoughts and ideas, and they began to learn how to express themselves to those with different experiences and backgrounds. There is also the hope that their stories can be used for advocacy, to stimulate more or better assistance to them or children in the same situation as they are.

But there is also the worry that discussing what can often be painful and frightening memories will lead to harm, further traumatization, revisiting horrors without the resources available to counsel the child, to work through a healing process. There is also the problem of disappointment when the interviewer with whom the child formed a connection and some degree of friendship leaves again, never to return.

"How sad I am that you do not think of me anymore," Barika wrote me in a letter just days after I left the camp in Tanzania where I had met him.

He was an orphan, like Keto, and at twelve years old was struggling to make a difference in his community. He performed in a theatrical group that demonstrated lessons about AIDS and violence and other social concerns to youths throughout the camp. He is one of the most admirable young men I have ever met, and I had no intention of forgetting him. He would leap about as we spoke, acting out his story, miming machine gun fire, smiling like a sadistic soldier as he burnt the village down, and then whimpering like the little children—himself among them—who had fled across the lake to Tanzania. Barika did not have many friends, due both to the stigma surrounding orphans and the more mundane reasons that adolescent overachievers everywhere have few friends: he liked to study and to read and to think about difficult things. He reached out to adults looking for the friendship that his peers did not provide. When I came along and took him seriously, listened and watched, asked questions and wrote down his answers as if he were the teacher and I the pupil, he felt, as he later told me, a sense of importance. His rage when I left—rage that continued in letters for over a year despite my best efforts to convince him I had not forgotten him—was understandable. All Barika knew was losing—his parents, his home, his friends. I was one more loss in a short life full of losses, and he did not want to forgive.

Anna Freud points out, from her research with children who survived the bombing of London during World War II, that children can generally cope with the day to day horrors of war—and even grow tired of them—"so long as it only threatens their lives, disturbs their material comfort" but when the war threatens to break up the family unit, it takes on far more serious significance. Psychological problems, Freud noted, were more prevalent in children who lost or were separated from their parents during the war than among those whose families survived intact. For those children the war was a chapter of their lives that could be closed for the most part. For orphans and separated youth, the war was the defining event in their young lives, the time when everything changed, when safety ended, when their place in the world was ripped from them, when they became alone. In her memoir, *First They Killed My Father*, Loung Ung writes about the multitude of horrors she suffered under the Khmer Rouge in Cambodia as a little girl, but the theme pervading the entire book, from the first page to the last, is the loss of her family.

My leaving Barika after about a week of knowing each other cannot compare to the loss of his family, but it was, perhaps, a reminder of those other losses, a reinforcement that he was on his own.

I sent him a copy of *The Little Prince* in French and English so he could practice and so he could read a little about another kid on his own, having adventures. I sent the gift because he loved to read. I sent the gift because he lived in a world filled with too little kindness. I sent the gift as a Band-Aid for the wound I had torn open, the kind of wound that never really heals.

I have tried to stay in touch with many of the children I have met, but the realities of war and displacement have made it difficult to keep track of the kids, to get messages to them or receive them again. Barika and I have stayed in touch a little, from time to time, though the distance and the vastly different worlds in which we live have limited our contact.

In that first meeting with Keto, I knew that the few hours we had together would probably be all there was. I wanted to connect with him very much, but I did not want to do him harm with that connection. No research was worth that. My desire for understanding did not outweigh his needs, not by a long shot. I had to tell him something in answer to his question, and it had to be true. You don't bullshit a child who has seen what he's seen, survived what he's survived.

"How will talking to you about the war help me to get shoes or more food or a blanket?" Keto scratched his chin and waited for an answer.

"It won't," I said.

The noises of the camp, of the world outside the tent filled the space between us. I heard laughing and loud conversations. The silence was awkward in the dim light between Keto, the translator, and me. The silence was long.

"It might not help you get shoes or more food or a blanket," I told him. "Talking to me won't get you those things." I paused and thought about what I was there to do, what I hoped would come out of it. He rested his chin in his hand and watched me speak, not understanding what I was saying until it was translated, but listening intently to the sound of my words.

"But if enough people hear your story, perhaps they can start a larger effort to help all the kids living in this camp. Your story could help put shoes on the children after you. The people who read it, perhaps they will want to help other children, perhaps you can teach them about what it is like so it can be made better here one day."

When I said this, I believed it. I cannot really imagine that Keto did. His face did not betray any reaction as my response was translated for him. It was not his first time being interviewed.

"I talked once with a *mzungu*"—a white man—"from the Methodist church. He interviewed me and he helped me, gave me a blanket. But it got stolen," he said.

I told him that I did not have any blankets to give, though he had not actually asked. Savvy Keto knew how to get me to volunteer what he wanted to know. This gift for people, for reading them, for getting adults to open up to him, was a gift that had served Keto well in the past, would serve him well in the future. Maybe it was this that kept him alive when the world around him fell to pieces.

We were eating granola bars while we talked, and I think he enjoyed that, perhaps saw it as fair barter for his story.

"I'd like to help other children," he said. I think he sensed my nervousness and wanted to make me feel more at ease. It was his turn to smile, to reassure me that he would talk, that he liked talking to visitors. It was embarrassing to be comforted by a child, a *victim* of war. I was supposed to be the expert here, providing the comfort, the support. My nervousness leveled the playing field for him, gave him a role to play. Keto did not want to be patronized; he wanted respect. He got it. He has it still.

"It's okay," he said and settled back in his chair and started talking again. He told his story without interruptions, except to let the translator speak or to nod when I seemed to understand something on my own because he had used a French word that I knew. Both of us liked those few moments of direct connection as he spoke, but otherwise, he spoke without much emotion and without many pauses. . . . It amazed me at the time, though I grew used to it over the years, how so many children who had been through unspeakable horrors could talk about the most disturbing things with little emotion. These were the facts of their lives. These were their stories.

"I came from Baraka," Keto said. He told his story, how he sat in school with his brother listening to the teacher recite the French lessons for the day: *je m'appelle, tu t'appelle, il s'appelle*. . . .

"When I went home, I didn't find my parents. My brother and I didn't know where my parents or grandparents were." They stood for a while in their empty home, calling for anyone they knew. With gunfire and flames around them, the two boys decided they must escape on their own. They made their way to the lake still clutching their schoolbooks to their chests. "They were our only possessions when we fled. I still have them after all these years.

"We went first to Fizi, a place near where I am from, and there we found crowds of people. There we found my father's brother. He said we should leave the Congo, but he wasn't prepared to flee. We were later told that that uncle, who did not come with us, had been burnt to death in his house."

In Fizi, the boys found their mother again. "I don't know what happened to my father—I have not seen him again— but my mother took us to Kibrizi, where refugees go when they get to Tanzania, and then we spent three years in the Nyaragusu refugee camp. Mom died in Nyaragusu. We heard lots of things . . . that mom died of AIDS, but I was young and didn't understand. People were scared to care for us; they thought that I had AIDS, so we stayed with another uncle, but he left at the repatriation, went back to the Congo."

"Excuse me," I interrupted. I looked at the translator. "Did he say 'repatriation'?"

"Yes, he did," my translator told me. "Some children learn all the words used by the UN, especially the unaccompanied minors. This one speaks very well, is very smart."

I turned back to Keto. I was amazed at the vocabulary that refugee life gave this kid. Words like "repatriation," "transit center," "food rationing," and "distribution." These terms are a reality for millions of the world's children, and they learn them in order to survive. They are magic words, words that open doors. Conflict creates a new vocabulary, and the dependence these children have on international aid teaches them to speak its language. I was reminded of the sisters, six and nine years old, that Anna Freud mentions in her writings on the Hampstead Nurseries during World War II. Walking down a London street after an air raid, the girls would look at houses and declare "Incendiary Bomb" or "High Explosive" based on the damage. It was not a morbid interest in the weapons of mass destruction around them that made these two British girls munitions experts, it was just the world they lived in. For Keto, his world was humanitarian policy.

Keto's story sounded not rehearsed but performed, as if he already knew that part of living as a refugee was telling your story to foreigners, the price of admission to refuge.

The asylum narrative is part biography, part myth, part plea, and part propaganda. It is how one person places himself, his terrible ordeals, in a larger context; how he makes the unreal real to those who can only imagine, how he becomes more than an individual suffering, part of a movement, a refugee, and in adding his story to the larger story of a people, of the displaced, he is simultaneously unique and not alone.

"Ready?" my translator asked. "He wants to know if he can continue."

"Um, yes," I said, flipping the page in my notebook and dropping my pen. He waited for me with his chin resting on his hand—his default position it seemed—while I got ready again. Keto was clearly the one in control here. "What did you do after your uncle left?"

"We stayed with his girlfriend. My older brother didn't like this woman, so he went back to the Congo, and when I was alone with her, she started hating me. That's when Christian Outreach transferred me to this camp."

In Congo, as in much of sub-Saharan Africa, AIDS is destroying many of the normal structures that ensure children are not abandoned. In the past, in Congolese society, when a child's parents died, the community took the child in and provided for him. The fear of AIDS, combined with intense poverty, disrupted this practice. Many children who lose relatives to AIDS are discriminated against, harassed, and often turned away. Besides the emotional turmoil of losing a parent, they have to face the hardships of abandonment, prejudice, and fear. And because of the war, soldiers moving around the country, taking many women, raping or sleeping with prostitutes, AIDS is spreading: a new kind of weapon in wars that are aimed at destroying civilian life. Many children do not have Keto's good fortune in finding someone to care for him. And many, like him, had to flee not just the fighting, but the stigma of the disease, often more devastating than the soldiers. Communities do not want the burden of an *unclean* child. A recent trend suggests many of these children in the eastern Congo are being accused of witchcraft and sent away. They often find themselves working as child laborers in mines, benefiting little from the adults who exploit them. They are outcasts, unwanted, unmissed, and their deaths are rarely noticed. Many such children have met their end in the bottom of abandoned diamond pits.

"In Lugufu, there was a man who knew me and knew that Mom had died. I stayed with him. He made sure I was studying, but I couldn't afford the school fees for secondary school, so I had to go back to primary school. I started cutting grass and collecting it for people to build with so I could have money to pay for school, so I would be in school one month and out of school the next, working. Then they lowered the fees, so I could go back to school all the time. So that is my story of how I crossed. The journey was hard, but I don't think I want to go back to Congo. I don't know that I have anything to go back to."

He doesn't.

The Democratic Republic of Congo should be a wealthy nation. The ground is rich in gold, copper, diamonds, zinc, and coltan (a mineral used in cell phones). As is the case in many developing countries around the world, the presence of natural resources is, for most of its citizens, a curse. If the elites can draw their wealth directly from the ground, what need do they have of taxation? And without taxation, what need do they have of the people? Their mandate comes from control of the ground, control that can only be gained, can only be held by bestowing favors and by force. The people stand in the way of access to the earth, to the riches beneath the surface. The people demand services, protection, a piece of the wealth under their feet. Power dragged from the earth has no love of the people; they are easier to deal with buried in it than standing on it. For this reason the history of the Congo has been a history of brutality. It has little to do with tribalism and "ancient ethnic hatreds," the oft-spouted phrase that hides the true nature of the conflict: wealth and power.

When the Congo was a Belgian colony at the dawn of the twentieth century, it was the personal property of King Leopold II. Using forced labor from the locals, he extracted a fortune from the vast territory, which is almost a hundred times the size of little Belgium. It is said in Congo that the streets of Brussels are paved with Congolese gold. The price for this gold was an estimated 10 million human beings dead, due to King Leopold's policy of destroying crops and villages to quell revolt, forcing men, women, and children into slavery, and the free use of capital punishment.

After independence from the Belgians in 1960, a U.S.-backed coup ousted the first democratically elected leader of the Congo, Patrice Lumumba, and placed the government in the hands of Joseph Désiré Mobutu in 1965. Mobutu was then the Army Chief of Staff. He was to rule for the next thirty-two years, stealing about 4 billion dollars from his country while many of the inhabitants remained some of the poorest people on earth. To describe his style of government, journalists used the term "kleptocracy."

He changed his name to the praise name Mobutu Sese Seko Kuku Ngbendu wa za Banga, and changed the name of his country to Zaire, a Portuguese corruption of the name of the country's largest river. He created "Mobutuism," a policy designed to drive colonial influence from Zaire. To this end, he outlawed the necktie and designed a new fashion statement, the *abacost*: a two-piece outfit of pants and a tunic worn with an ascot. He was a master of the politics of the ground. He bestowed favors on his allies, access to the minerals and metals in the soil, and a share of his largess. Foreign companies and local Big Men benefited greatly from his favor, even as the people starved. Much like Marhsal Tito in Yugoslavia, he suppressed ethnic nationalism when it threatened him, and divided people along ethnic lines when it served his interests. The army wasn't paid—he told them to fend for themselves with their weapons and take what they needed from the people.

His rule grew more and more precarious. In 1995, the parliament passed a referendum that would strip Mobutu of all real power and leave him as a figurehead to create a smooth transition to democracy. It was one of the first and boldest democratic moves in Zaire since the coup that ousted Patrice Lumumba. Mobutu ignored the referendum, as did the rest of the world. But the frustrated political opposition to his rule turned into a military opposition soon enough.

With the backing of Rwanda, Uganda, and Burundi, the Marxist guerilla Laurent Kabila (who had fought briefly with Che Guevera during the late sixties), formed the Alliance of Democratic Forces for the Liberation of Congo-Zaire (ADFL) and marched on Kinshasa. He ousted Mobutu in 1997, and the man once known as the Father of the Nation became a refugee himself. Mobutu died in Rabat, Morocco, shortly after his exile began. He is buried there in a cemetery now, ironically, given Mobutu's legacy in Africa, as "Pax."

Laurent Kabila's takeover set the stage for the start of what would be known as Africa's world war. After the fall of Mobutu, Kabila changed the name of the country to the Democratic Republic of Congo. He had come to power with the support of neighboring Rwanda, but resentment grew against Rwandan power in the Congo, and against his support of the Banyamulenge, who were Congolese Tutsis, often labled as Rwandan themselves. In order to shore up his political gains, Kabila turned against the Rwandan and Ugandan governments that had helped him invade the country and overthrow Mobutu. The Banyamulenge in the east rioted. Both Rwanda and Uganda invaded. Other nations jumped at the opportunity to exploit the abundant natural resources of the new "Republic" and rushed to Kabila's aid, sending in their own armies. In 1998, all-out war began in the Congo.

At any point there were five armies of other nations fighting, as well as countless local militias, like the Mayi Mayi (which consisted of at least twenty different factions), and the genocidal *interhamwe* from Rwanda. Rwanda's legitimate army was fighting the *interhamwe* in the eastern Congo, as well as Kabila's government based in Kinshasa, and extracting a wealth of diamonds from the land to finance the war. They also backed the rebel group RCD-Goma, who controlled most of the eastern half of the country.

The war in the Congo was declared over in 2003. Four rebel leaders became vice-presidents. At the time of writing, fighting continues in the east, displacing thousands more people, some of whom had been hopeful that peace might finally come to their country with the end of the war and the first democratic elections since 1960. Azarias Ruberwa, the former head of RCD-Goma and now one of the vice-presidents (and a thwarted presidential hopeful with an army at his beck and call), suspended his party's participation in the Kinshasa-based government for four days. Ethnic conflict between the Banyamulenge (Congolese Tutsis) and other ethnic groups has flared, resulting in massacres, widespread human rights abuses, and continued violence. Militias and bandits still terrorize much of the Ituri and North Kivu districts and for many Congolese the war has never ended.

The war in the Congo killed nearly 4 million people as a direct result of violence or, far more commonly, due to malnutrition and disease exacerbated by the conflict. The International Rescue Committee reported that between 1998 and 2004, around 1,200 people died every day because of the war. That death toll is equivalent to three 9/11 terrorist attacks per week.

While young children are the most vulnerable—one in

four die before reaching five years old—it is adolescents who are the most susceptible to forced recruitment as soldiers, sexual exploitation, and exploitation of their labor. In short, it is adolescents who are most at risk for violent deaths. With their parents often unable to support them, adolescents in armed conflict are more likely to be sent from home to find work in the cities or to take on the burden of supporting the family themselves. Not yet adults, they are no longer nurtured as children. They have neither the protection of the young nor the rights of the grown.

Keto was happy to show me his self-confidence, his ability to manage for himself during the violence and instability in the Congo and the camp. He figured out how to pay his own school fees and aid in supporting the man who had taken him in. He is a central figure in the economic survival of his caretaker as well as himself. He does not have, nor does he seem to want, a passive role in his well-being. He would like assistance, but he knows how to bargain to get it: telling *mzungu* researchers his story, for example.

I heard similar stories to Keto's throughout my time in the Congolese camps in Tanzania.

"One day the soldiers came," Michael told me. Michael, who was fifteen, fled the Congo almost two years before I met him. He, like Keto, lived near Baraka in the area of Fizi, where he worked with his father running a table in the market. He was extremely well dressed for what I had envisioned an African refugee would look like, especially an orphan. He had on a clean blue Oxford shirt and long khaki pants. He also had sneakers that would be considered nice by any standard, not particularly coated in the thick red dust that covered pretty much everything else, as if he had cleaned them moments before I arrived to meet him. He had indeed done so.

He told me that he was suffering and that he lived on his own with boys his age who had also lost their parents, though he did not get along very well with them. He said they stole his clothes.

"I borrowed these clothes to come here and meet you," he told me. "One day, when I was bathing, I came back to get my clothes and the shirt I have was ripped. All my other clothes were stolen."

Michael sat very straight in his chair and smiled when he gave my hand a firm shake, like a businessman closing a deal. He was trying very hard to be like his father, who was a businessman. He used to travel with him, wheeling and dealing, he said in English: "Doing business." If he had money, he told me, he would start buying and selling, traveling around carrying on the business for his dad, whom he still wants very much to make proud.

"I was in the back room when the rebels came," he said. The rebels burst into his house, knowing his father was a businessman and would have money. The burst in through the front door armed with machetes and rifles. "That's when I saw my mother and father killed, and all I could do was climb out the window."

He scratched the back of his head and looked at the floor. I was about to speak, to help him move from this painful memory. He was fidgeting and quivering slightly at the lip. Then he sat up straight again and met my eyes dead on. He was pulling himself together, not wanting to stop the interview.

"It was chaos. I was running and everyone around me was running and when I got to the shore of the lake, I realized I had no money."

Standing at the side of the lake, young Michael started crying. Around him the world had erupted into violence. Moments earlier he had seen his parents killed. He could not go further on his own. If he stayed where he was he would either be forced into the army or die. Maybe both, in that order.

"The man who had a boat saw me and took pity on me," he said. "He said I could cross if I bailed the water out of the boat. So I got on and bailed water the whole time we crossed the lake. The boat was so crowded, and everyone was upset. When I first got to Kibrizi [the UN reception center] and saw the green plastic sheeting and all the people I didn't know, all I thought of was my parents. A car took us to the camp, and now I live here. I have no clothes of my own and no money to do anything. I think about my father. I go to school. I play football, but there aren't enough balls. I am a good striker. I win a lot of games."

Michael did not seem to get along with others the way Keto did. Usually, when I finished talking to a child in a refugee camp, a crowd would gather around that boy or girl, a multitude of other kids and several adults as well, wanting to find out what happened, what the *mzungu* wanted to talk about, what he had to say. Information is a valuable commodity in a refugee camp. When I walked, I tended to have a tail of about ten children behind me. When I walked with Michael, no children followed. One or two would come up and ask him a question to which he gave a brief answer, and then the other kids would look at me and the translator for a moment and walk away. No crowds gathered to Michael when we were done speaking. He walked off alone.

Keto used his considerable skills at reading people and an amazing amount of energy to manage his emotional and physical survival. Helping the "old man" he lives with, I believe, provided him with some of the strength he needed to deal with stresses he experienced and continues to experience.

Alienated from his peers, Michael held on to the image of his father for support. He told me it made him sad to think of, but that it also gave him a goal. He held on to his past as a source of hope and enjoyed talking about the ways his father would trade and make deals, would drive around on the motorcycle he had, selling goods. Michael was not at all content with his situation. He didn't go to school anymore, and he didn't like living with the other boys his age. He wanted to get out into the adult world, not be regarded as a child anymore. I think he was frustrated with the assistance programs that treated him like a child, giving him no say in his own future.

"I escaped from the Congo. I know how to travel. I would like to travel and see the world," he said. "Not stay here."

Then there was Melanie, a girl I met the same day I met Michael. She had just been playing jump rope at the well when we met, and her red dress was splotched with dust. Her tiny hands fluttered like moths while we talked, always moving or picking at something.

She has few memories of her past from which to draw inspiration (at least that she told me about) and her future is highly uncertain. "They tell me I am thirteen, but I do not know," she said. She lives with her teacher, the wife of a man who rescued her from the fighting in the Congo.

"I do not know where I come from. When we fled the fighting in Congo, I missed the truck with my mother on it. The Congolese soldiers put her on a truck. I do not know where she is now. She is Rwandan. I got lost and a man found me in the brush. He took me with him to Tanzania, but there were guns."

"Guns?" I asked, looking at her drawing which was filled with weapons (Figure 5).

"I saw the guns when they came to kill the man I was with. All the things, guns and spears. They wanted to kill the man because he was a soldier. We hid for two days under a bridge that had a roadblock on it. Then he paid for me to take a boat into Tanzania. I was happy to get to Tanzania because of the war. When I arrived I felt safe. They told me not to speak Rwandan, because of the militia, even in the camp. My teacher taught me to speak Swahili, so now I do not speak Rwandan with anyone. It is safer for me. I feel safe now, in the camp, most of the time. My teacher says she will help me find my mother when the war is over and we go back to Congo. I will let my mother decide if I should stay with her or with my teacher."

Melanie is not simply vulnerable because she is young or because she is a girl. Her ethnic status—Rwandan, as she said, meaning Tutsi—makes her more vulnerable. Her teacher's warnings reveal that even the act of speaking could be dangerous for her. She must deny her own language in order to survive. Her safety, even outside the war zone, is linked directly to the political situation in the Congo. Young girls from different ethnic groups have very different experiences in the refugee camp, though only through spending time getting to know Melanie would someone realize how different her needs might be from other little girls, how different the threats against her.

The Rwandan government is the principal power backing the rebel government of the eastern Congo, the RCD. Their enemies, such as the Mayi Mayi, the *interhamwe*, and the recognized government in Kinshasa, have created a fear of the Rwandans, of Tutsis, that runs deep. Through their own human rights abuses, their actions without regard for Congolese civilians, and their exploitation of the Congo's resources, Rwanda's army and the RCD have helped in the creation of this fear.

"The war is caused by Rwandan soldiers," said Robert, a young boy living in a shelter for street children in Bukavu in the eastern Congo, just across Lake Tanganyika from the refugee camps in Tanzania.

"The Rwandans kill people. They massacre. I've seen it myself, you know. I've seen someone sell a cow and everybody knows he sold the cow and the soldiers came and demanded money from him. If he didn't give them money, they'd cut his throat. I could find a Rwandan who would cut a throat for $100. I could find him with you right now, if you want to."

I declined the boy's offer.

The fear of the Rwandans illustrated to me just how difficult it was for young people to form their own identities when the situation around them dictates so many terms. The young have an advantage over adults, because they can be more adaptable to new settings, but they have the added pressure of adapting to a new society at the same time as becoming socialized within the family, sometimes not their own families. It is hard enough for an adolescent to define him- or herself in peacetime, managing the expectations of family and society while trying to define oneself. Add to that armed conflict and ethnic strife, and one can imagine the challenges adolescents like Melanie face. This makes her cheerful manner all the more remarkable.

Melanie was forced to change her cultural heritage, abandon her language, and has formed an attachment to her teacher. If she were to embrace her past, embrace the memory of her mother and her mother's people, she would be in danger. I could not determine with which army the husband of this teacher fought, but it is possible that the dangers of failure to adapt, of forming new bonds with this teacher, are more than psychological and could come from within the family that cares for her. Luckily, it seemed, Melanie enjoyed her relationship with her teacher, who genuinely seemed to care for her. At the time we met, she did not seem to be suffering from the sense of loneliness that plagues many unaccompanied minors. Perhaps because of the effectiveness of her assimilation, she managed to make friends and find supportive adults, unlike another Tutsi boy I met, Justin.

Justin, fourteen years old, lived in Kigali, the capital city of Rwanda. Unlike Melanie, he often felt lonely, isolated, and afraid. He was a tall, gawky kid who liked to study and read when he had the chance. He did not have many friends. Justin fled to Congo with his mother during the genocide in 1994. His father didn't make it out, he told me.

"We lived in Congo for a while, then the war there started. People were killed; people remained behind, desperate. That is where my mother was killed."

I have to assume from the time he fled that Justin is from the Tutsi ethnic group. This is not a subject we would discuss openly in this refugee camp. It would not be safe.

In April 1994 Rwanda President Juvénal Habyarimana was assassinated. Within a few hours, the Hutu Power government, built on a platform of ethnic superiority to Tutsi minority, accused Tutsi extremists of orchestrating the assassination. They called on the population to kill every Tutsi in the country as a matter of national security. Though the outbreak of violence seemed spontaneous to outsiders, General Roméo Dallaire, the UN force commander in Rwanda at the time, observed, the Hutu Power Movement had been preparing the mass slaughter for months. As early as February, teachers had been recording the ethnicity of their pupils, even though children were not required to carry ID cards. It would not become apparent to the general what had really been going on until teachers began murdering their Tutsi pupils.

On the night of April 6, the Hutu ethnic majority in Rwanda began a campaign of extermination against the Tutsi minority and against moderate Hutus. In around a hundred days, or three months, more than 800,000 people were killed. Most of the violence was committed with farming implements, hoes and machetes, mostly by bands of young men, the *interhamwe*, which translates as *those who attack together*. The *interhamwe* later fled to the Congo, providing the pretext for Rwanda's invasion and occupation.

There are accounts from the spring and summer of 1994 of neighbors killing each other, of priests and nuns killing their Tutsi parishioners or moderate Hutus, of mixed families turning the blades on their Tutsi wives or husbands or in-laws, caught up in the genocidal fever. Thousands of Tutsis and Hutus opposed to the genocide fled to the Congo and Tanzania to escape the killing. Paul Kagame, an exiled Tutsi soldier, led an attack on Rwanda from Uganda and took over the country. He stopped the genocide by the end of the summer. The *interhamwe* and the former Hutu Power politicians—the architects of the genocide—along with hundreds of thousands of innocent Hutus, fled from Kagame's army into Mobutu's Zaire. The images of this exodus, captured in vivid photographs by Sabastião Salgado, seem like a relic of another time, biblical or medieval, but certainly not the end of the twentieth century, certainly not 1994.

When conflict erupted in Congo (then Zaire) in 1998 after the ouster of Mobutu, many Congolese Tutsis or Banyamulenge, who were not Rwandan, became targets of the Mayi Mayi local defense forces or of government forces. Hutus from Rwanda, whether they were involved in the killings or not, became targets for RCD-Goma and the Rwandan army.

Still unsafe in the Congo, Rwandans of both ethnic groups fled once more, across the lake to Tanzania. Children, like Justin and Melanie have spent most of their lives without a homeland, without permanent homes. The dangers of ethnicity, of ethnic nationalism, the dangers of hate have chased them from their homes and followed them into the camps.

Justin told me that he saw his mother killed; he was hiding and watched it happen.

"One day the soldiers came and they cut my mother. They killed my mother with the big knives they had. I tried not to look, but I heard the noises they made and she made. Not loud noises, but I remember them.

"I ran away, and while I was running, I hurt myself. I met a Banyamulenge man." The Banyamulenge are another ethnic group in East Africa, Congolese brethren of the Tutsi in Rwanda. "I told him my problems and cried to him. He was kind and he helped me get to Tanzania. The family I lived with first, they abused me. They took my food and blankets and were very cruel. I do not know if it is because I am an orphan or because I am Rwandan. I do not know why. I was moved by the Red Cross and live here now. It is very bad. I cry every day when I get home from school. I think about my mother and no one comes to comfort me." By this point his eyes welled with tears. "I do not know how I will get over this. It would be better just to forget."

Without any ties to his culture or his family, Justin feels adrift. He is lonely, he says, but he is beginning to feel better. "I am learning to forget.

"I like to go to school, though there are not enough books." When I try to focus on those aspects of Justin's life that he finds positive, that are making him feel better, he does not hesitate to answer: "I went to a training for children about rights."

CORD, the organization that helps provide for the unaccompanied minors in Lugufu camp, has given adolescents training in children's rights. This kind of involvement in his own well-being has given him motivation to wipe the tears from his eyes, go outside, and get involved with the world he lives in. As he speaks about rights, showing me his drawings on which he has written various empowering statements from the Declaration on the Rights of the Child, he becomes more animated, visibly more confident and eager to talk. He does not avert his eyes to look at the ground as he had for most of our conversation until that point.

"I learned that children have the right to go to school." He shows me his drawing of a boy walking toward a church. Written in Swahili above it is, "The child has the right to do all kinds of work and go to school" (Figure 6).

"School will help me get a good job and become a professional. I would like to live in an urban area again. Here, the environment is very bad. Sometimes people don't even use the toilets. And when you get sick, it is a long walk to the hospital and then, sometimes, you can't get anyone to help you."

Justin's concerns about public health and cleanliness, his concerns about school resources, were pressing on him. He was aware that schooling was a way to secure his future, one of his "rights," and that he was in danger of disease from the poor hygienic conditions in the camp. These stresses weighed on his mind a great deal—he brought them up or alluded to them during our talk several times, expressing frustration and once, nearly cried when discussing the uncleanliness. He felt helpless against these things and had no one to whom he could turn. Though he understood the rights he and all children should have, he could do little to realize them, and that might have contributed to his sadness, the realization of just how much he was at the mercy of forces much greater than he was, entire governments and armies and institutions that controlled his fate. A heavy burden of awareness for anyone, let alone a fourteen-year-old orphan.

I liked Justin, though I'm not sure why. He was not as well-spoken as Keto, nor as eager to impress as Michael, nor as cheerful and buoyant as Melanie. He was charismatic, and I wanted him to feel his own worth. With his interest in children's rights evident, we talked about my project of researching the lives of young people affected by war. He liked the idea of being an ambassador for young people in situations like his.

"What would you tell someone your age who has never been in a refugee camp so that he could understand what it is like?" I asked him. Justin thought for a moment, choosing his words carefully.

"I would like to tell my name so that he could know me," he answered. "I would tell him that living in the camp is very bad. I think about going home, but who will I go back to? Everyone is dead. If I talk to this boy who has never been in a refugee camp I would be happy. I want to find children with hope."

*These journeys that children are forced to make are not con-*fined to Africa. Right now, there are an estimated 20 million children uprooted from their homes around the world, living either as refugees, "migrants," or internally displaced persons.

"There were many hardships on the journey [from Burma]," Siha said. He was sitting on the floor of the largest room in his little house in Thailand, in a city where many illegal Burmese migrants sought safety. He wore a soccer jersey and black running shorts and poked his tongue out in concentration as he drew his pictures, like eleven-year-olds I had met in other parts of the world. He lived with his aunt, his cousin, and his mother, though his mother was away for a few weeks at the time we met.

"We walked for two days, and it was raining the whole time, and then we rode horses, but my mother and aunt walked. And the river was flooded. We rode with buffalo and cows on a boat and it was very hard. I was afraid to leave home, but I was with my mother so it was okay. Everything was different here. The place to sleep and the place to live were different. We did not know where we would eat or what we would eat. In Burma, my grandmother would send me to the market for her and it was a very long way. I remember going there and walking far from home to buy different things. Here we did not know what things we would have."

Siha is considered a migrant because, as a member of the Shan ethnic group, he is not eligible for refugee status in Thailand.

During his journey, Siha had the protection of his family, his mother taking care of things, making sure the children could ride horses instead of walking. As psychiatrists Joseph Westermeyer and Karen Wahmanholm observed in their work with refugee children, fleeing can seem like an adventure if children have a parent or parents insuring continuity and taking responsibility for their survival and well-being. Culturally, the differences in what constitutes childhood affect the way the young experience flight into exile, as do the differences in the wars being fought. In Congo, with the ravages of AIDS and the protracted intensity of the fighting, societal norms have broken down to such a degree that family structures become unraveled and few people have the resources, either emotional or material, to support children who are not their own. Additionally, it is not unusual for young boys to have responsibilities outside the home or for young girls to take care of their siblings. Unaccompanied minors are much more common in the Congo and in refugee camps in East Africa than in the Burmese communities in Thailand. This could also be due to the fact that on the journey out of Burma into Thailand there are several checkpoints controlled by one or another army, dense jungles filled with land mines and armed patrols, and young people on their own simply do not survive.

"It was hard to cross the border. There are robbers, Mon soldiers, Burmese soldiers, Karen soldiers, all wanting money. We had to pay many times at many checkpoints. It was dangerous," said Nicholas, an eleven-year-old boy from the Karen ethnic group.

On their own, children cannot pay the bribes; perhaps they join or are forced to join the soldiers; perhaps they are turned back by the soldiers and sent home. The Thai authorities regularly round up the Burmese refugees (they are seen as illegal migrants), and send them back to the border areas. Unaccompanied minors would be easy targets for these roundups. They would also be easy prey for the flourishing sex industry in Thailand. On the streets of Bangkok, one can see countless young male and female prostitutes. Due to all these conditions, it is harder to find the Burmese youth who are living on their own. Neither the pimps nor the Thai police nor the Burmese military or rebel groups are inclined to give researchers like me access to the children they control. Though these children may exist in Thailand, I would not have the opportunity to meet them.

Among the Burmese children I met, most of whom had at least one member of their family with them, the troubles of the journey, of the violence witnessed or experienced, were mitigated by the support they received.

"Sometimes I don't sleep well, and my mother comes to me. I tell her I'm having bad dreams and she tells me it's okay. We're here now and we are safe. But I don't always feel safe," Nicholas confided in me. We were in a city near the Thai-Burma border, a place where police corruption is rampant and smuggling flourishes in diamonds, drugs, weapons, and people. Burmese children are particularly vulnerable, and these stresses bother Nicholas as much as the memories of his village.

"One day the SPDC"—State Peace and Development Council, the name of the military junta—"came and burnt my village, so I wanted to draw this," he said, showing me the picture he produced when I asked a group of school kids to draw anything they liked. "I don't know why they did it, burnt my village. I ran with my family into the mountains and crossed into Thailand. The army would have arrested us if they'd caught us trying to leave, but we snuck out through a secret way."

His drawing captivated me.

Nailed to a cross, a young man cries out as soldiers fill his body with bullets. To the right, a soldier climbs a flagpole and takes down the unmarked flag. Bodies fall from the sky, dropping from an exploded airplane. On the far left, easy to miss at first glance, is a little form in purple, a boy hiding behind a tree (Figure 7).

Nicholas's blank flag, the flag of the defeated, shows a keen awareness of his situation. He is an illegal migrant hiding in Thailand, unable to attain legal refugee status and clearly unable to return to his homeland. He doesn't speak Thai. He has no nation. The blank flag is central to his picture. Even as his little purple form—he shrugs when I ask if that is him—witnesses terrible violence, he also witnesses the political struggle occurring around him. He cannot verbalize the politics of the fighting, but he has a sense of them: it has something to do with that flag; the reason he and his parents fled is connected to that flag.

Violent forced migrations, political struggles robbing children of their homelands, are not unique to the developing world. In the Serbian province of Kosovo, in the last years of the twentieth century, a policy of ethnic cleansing filled the roads with terrified people running for their lives and killed hundreds of fathers, brothers, and uncles. The violence displaced nearly one million people.

One of those was a young girl named Nora from the village of Zahaq. She was about eight years old at the time she had to flee the country.

It was in May. It was a sunny day. I was playing in front of my house where there were fruit trees. Serbs were hiding behind those trees. They came into our yard and asked for money and jewelry, asked where the men were. My father was in Albania fighting with the KLA [the Kosovo Liberation Army]. My grandfather was at my uncle's house. So I lied to them. I told them they went out for cigarettes. They believed me, but they asked my mom about where my dad was.

She told them that he had had an accident with a train, and they believed that too. I went with [my mom's uncle] and my mom, but the Serbs caught us. They put a knife on my neck. They wanted to rob us and they saw my mother's wedding ring and they told her to give it to them. It was hard to get off. She struggled with it. They said, "Hurry up or we'll just cut off your finger!" But she got it off and they let us go. They still searched the house to see if anyone was hiding there, then they made us leave. We went to my uncle's, but couldn't stay there, so we went to my grandfather's house, and he told us to leave the country. He said, "I'm old"—he's eighty years old, or was then anyway. He said, "I'm old and tired. I can't come with you. My legs won't carry me on this journey." He hid himself between some trees.

They took one of her uncles, though. They shot him with a silencer and dragged him inside his house and burnt him there. They wrapped him in a blanket so he would burn easier. "We knew it was him later by how we was wrapped and that his face was not burned. I saw him after the house burnt down."

As she told me this story, I felt overwhelmed. So much could happen to a little girl on a sunny day in May. It happened near her school where we were sitting five years later. There were three other young people from the village with us, all of whom had also suffered terribly at the hands of the Serb paramilitaries. The others wanted to tell their stories too. Nora told them to hold on. Her frightful tale was not yet done and she wanted to make sure the narrative was complete.

"We went to our neighbor's home and lots of people were gathered there listening to the news. The paramilitaries came and beat us with rifle butts and clubs and their fists and told us to say good-bye to each other because they were going to kill us soon. We stayed there two more weeks, though, and the Serbs didn't come back. We stayed two more weeks until some other Serbs came, put us in a line, and made us leave."

The children began to describe this line, which, at the time, they and all their families joined. It was a line of people and vehicles starting in the center of town and directed to head out of Kosovo towards Montenegro. There were buses on the road. Those who had cars loaded them with their families and whatever property they could fit. People rode on tractors and horse-drawn carts. People also walked. It was Gibbon's highway "crowded with a trembling multitude." It was along this line that the next wave of horrors occurred.

"Army men in black masks stopped us," Nora said. "They took some men from the line, who disappeared."

Mark, who had been eager to interject, finally cut in.

"They took my father when we were in this line. They took him from right in front of me and two of my first cousins too, and shot them. But my father survived. He lay under the bodies until they were gone."

"They tried to take my father," another boy, Karl, said. "They didn't though. The truck driver that was taking us, turned around."

"I was with Karl on the same truck," Valerie added. She brushed her long blond hair from her face. "They took my father and five of my uncles. They killed them."

Human Rights Watch reports that nineteen people were killed in Zahaq on May 14, 1999.

"They killed my father later," Karl said. "When the Serbs were pulling out [after three months of NATO air strikes on Serb positions], a yellow Mercedes came into the village, and the men in it shot him."

The retention of details amazed me, as did the way these stories played out for children of all social classes. In Africa, it had mostly been the poor who were forced to flee, as most of the fighting took place in or around villages. Wealthier citizens stayed in the cities, hired protection, or left the conflict areas altogether with resources to pay bribes and avoid refugee camps. The youths I met in the Balkans, in the villages of Zahaq and Lubeniq and Pavlan, were poor even before the war, and Kosovo remains the poorest province in the former Yugoslavia. But in this war the wealthy also had to flee and children of well-to-do families were no more protected from the terror of expulsion than anyone else.

Eric, an energetic twelve-year-old, and his fifteen-year-old sister Alice, are the children of a wealthy Kosovar Albanian businessman.

"In the afternoon on the Saturday after the NATO bombing started," Alice said, "the Serbs came to our house."

Eric interrupted her with a great deal of eagerness:

"Captain Death came to our house." The children didn't know his real name. But according to Human Rights Watch, Captain "Death" ("Mrtvi") was a known paramilitary and criminal leader in the city of Peja (Peč, in Serbian). His real name was Nebojsa Minic. He was directly implicated in the murder of six family members on June 12, 1999, but the siblings tell me: "He killed sixty-eight people." They repeated the number: "Sixty-eight." When I asked how they knew, the kids said they heard it somewhere. They added, "He killed our neighbor the same day he came to our house."

"When he came with the paramilitaries," Alice continued, "the minute they came, my mom gave them ten thousand Deutschmarks, rolled in a tube like a cigarette. She told him it was for all the other houses, too. But he didn't care. He burned our neighbor's house with twenty-four people hiding inside."

The people got out of the house, and saw that the paramilitaries had killed one of their other neighbors. They were looking for all the men.

"Captain Death came back a second time. He took my mother's hand and said, 'You survived because you could pay.' My mother doesn't remember this because she's traumatized," Eric told me matter-of-factly.

"He came back a third time and asked to see the basement because he thought someone might be hiding there. People had been hiding in other basements, and he wanted to find my father because he was rich. He was marked for execution."

The children understood that it was not just that they were Albanian that put them at risk, but that they were wealthy. They knew that they had survived because they had money—their mother's bribe to the captain—but they understood that their wealth might also endanger them.

"When they came back that third time, my dad and my uncle were on their way to hide on the roof. When Captain Death and the other Serbs came through the door, my aunt was closing the attic, so my mom came clomping down the stairs, making as much noise as she could to hide the sounds upstairs. Death asked my mom, 'Where are you going?' 'Leaving,' she said. 'Oh no,' Death said. 'Not yet.'

"Then he lined us up.

"Right there," Eric said, pointing out of the living room into the foyer at the base of the stairs. "There were my cousins there too. Thirteen people he lined up."

Captain Death put a gun in Eric's mother's back, and she told him that she had a Mercedes in the garage. He could take it if he wanted. Captain Death replied that he already had some Mercedes and didn't need another one, but that they would go inspect the garage together. He shoved her there with the gun in her back, in case someone was hiding to ambush him.

"Once he saw it was all clear, he left again," Eric told me. He did not know how long they spent in the garage, and I did not want to push the question. Whatever happened in that garage was none of my business. I knew what the paramilitaries did to Albanian women."

Some time passed in our conversation. This was not an uninterrupted narrative that Eric gave me. I have edited down a conversation over several hours with the usual asides, meanderings, even a discussion about a popular Spanish soap opera. In that time Alice left the room to go watch television and gossip with her cousin. Eric continued.

"The people from the burning basement came out and saw our dead neighbor. They were screaming and yelling. At the same time, we heard gunfire somewhere. We thought we were all dead. My oldest sister went out to check what had happened. She saw that our neighbor was dead, shot. She called out and everyone came and saw. They wrapped him in a blanket and buried him in the yard.

"We stayed up all night. The men patrolled the area. Early in the morning the next day, we got some things to leave. My father and uncle argued about the Mercedes. My father wanted to leave it, my uncle said we needed another car. So he drove it and my father disguised himself and drove in the Peugeot.

During the drive to Montenegro we were stopped many times by the Black Gloves—a paramilitary group, mostly armed volunteer thugs and criminals under the loose command of the military—the unit wearing ski masks, the executioners, poked my father with their guns and demanded money. He was using a fake name until we got to Tirana, in Albania."

Eric's memory for the details is astounding. He also remembers when there was artillery fire near their house early in the war, and his mother picked him up to carry him, she slipped, and he fell. His father carried him the rest of the way. Eric thought it was important for me to hear this story too, to have a complete picture of his experience. He wanted to make sure I knew as many details as possible. He searched out and showed me a photograph of the neighbor who had been shot. I should see it, he said, and I could keep it if I needed to. I told him I didn't, but I wondered why he was so insistent on every detail, on remembering events so precisely. He had been correcting his sister when she spoke, which might have been why she left the room.

Other children showed the same attention to detail.

In the village Lubeniq, a fourteen-year-old boy, Leo, told of walking with his mother and siblings to Albania.

"The police came early in the morning, at six a.m., and made the whole village come out of our houses. They sent the women and children to Albania. They made the men stay. If we stopped walking to rest, the Serbs would say, 'If you don't get up in five minutes, we'll kill you.' They burned all the identification papers they found so we could not come back. When we left, I heard shooting. They killed eighty-four or eighty-six men."

Human Rights Watch reports that "between March 24 and June 10, more than eighty villagers were killed by Serbian forces."

Leo wanted to be very precise about the numbers—and his accuracy impressed me. The numbers must have come from adults around him, if not directly told to him then from his own active assimilation of information he overheard. Most of the children I met in Kosovo had this same impulse, get the numbers right, get as much information as possible about what happened, tell the names of the dead. It seemed to be a trend: The countryside is filled with memorials to fallen KLA soldiers, to civilians massacred. At a school in the mountains near the Montenegro border, a plaque commemorates an eighth grader who was killed in a mortar attack. The students I met, even the ones who were very young at the time of the war, knew his name and what had happened to him. Telling their story, for these young people, for this nation, seemed a way to validate their past, to prove they could not be expelled or exterminated. The story of individuals and the story of the people as a whole were bound together.

"There were nine survivors that day," Leo told me. "They were wounded, the rest were executed. Only twenty of the bodies have been recovered."

Leo, Karl, Valerie, and Nora, as well as many of the other youths I met, were bothered by the bodies that had not been buried, even if they were not the bodies of their family or friends. The unburied, missing bodies are a constant reminder of the horror these communities experienced. The wounds heal over time, but without the ability to properly mourn their dead, the scab leaves some painful grit underneath that never stops itching, will never stop until the missing dead can be put to rest and proper healing can begin. The stories they tell, five years after the war, are fresh and vivid in detail. I believe this is due, in part, to the fact that the story is not complete. Kosovo still does not have its independence from Serbia, though the UN administers it as a separate territory and NATO forces defend its borders with Serbia. No war criminals have been indicted in connection with the massacres or "cleansings" in Kosovo. This leaves the young feeling uneasy about the future. Their vivid memories are not "flashbulb memories," moments burnt into their minds the way some Americans remember where they were when Kennedy was shot or what they felt, saw, and heard on the morning of September 11, 2001. They are survivors' narratives; in a way, reminders that they lived through such things once, and that they could survive it again. They are, in all their horrific detail, celebrations of life for these young people, continuity of their experience, assurance that though some have died others go on. The children take the stories of their people forward.

Telling their stories helps to create a sense of community. The children in Zahaq were eager to tell their stories, supported each other's stories, asked each other questions to learn more about the escape from Kosovo, and filled in the gaps in each other's memories.

Barika, the Congolese to whom I gave a copy of *The Little Prince*, participated in a performance group that put on scenes about refugee life, that encouraged the youth to write poems about their experiences, and performed pieces about their history. Continuity of place is disrupted when the young are forced to flee their homes, but by sharing their stories they can maintain the narrative of their lives, both as individuals and as communities. Inviting willing young people into the process of telling can help them to heal, can help them to integrate their own suffering into a larger picture and, perhaps, combat the isolation so many of the young survivors of violence feel. Of course, the nature of this narrative matters too. The story can stir nationalism and ethnic hatred as much as it promotes psychological healing, as the Kosovo Serbs and Albanians taught me.

"Everyone was in the same position after the war," Nora told me toward the end of our discussion in Zahaq. We had left the classroom in which we had been speaking, and the five of us walked around the school. They wanted me to see the ways in which their building was broken. They were very concerned that I understand the present conditions, not just talk about the wars of the past.

"No one could talk about what happened," she said. "There was too much hurt. Everyone was in shock."

"How can I explain it to you?" Karl interrupted, looking around at the schoolyard, where a group of children were playing soccer and others were laughing and staring at us. "One day I lost my father and my grandparents." He looked at me in silence.

"But life continues," Valerie added. The others concurred. "Life continues."

THREE

"We Can't Stay Here"

*Migrants and Refugees
in Hiding*

Siha suggested that maybe I didn't see the fear.

"The children don't say it, but their parents are afraid and they are afraid." He had stopped drawing a picture of himself flying a jet in order to tell me about the fear that permeates the lives of the Burmese, then he returned to his paper. We were sitting on the floor of his aunt's house. As I had noticed was his habit, he poked his tongue out of the side of his mouth when he drew. His aunt told me that he would rather be playing video games than anything else.

Siha denied it instantly. "My favorite game is soccer. I love to play sports. I don't just play video games." He feigned anger at the false accusations against him, but smiled widely at his aunt. Everyone in the room laughed loudly, Siha, his aunt, my translator, and me.

Siha's aunt sat in her chair and listened to our conversation. I sensed she was something more than curious about the interview. When I suggested we play soccer for a bit to pass the time, the aunt urged us to stay inside, smiling all the while, pointing at the drawing material I had. She wasn't suggesting we stay inside; she was demanding it as politely as possible.

Times were dangerous for the Burmese migrants living in Thailand, and the attention brought by a foreigner could mean harassment, prison, or deportation back to Burma. Playing soccer with me was not an option. Lying low demanded keeping discreet company, even for a little boy in soccer shorts. As the only foreigner in this part of town, far from any plausible tourist attraction, my visit demanded a level of secrecy.

Siha's aunt sat nearest the door to the sparsely furnished room and glanced out the window over my shoulder. The shades were drawn, and she had to brush them aside with her hand to look outside. Her own son, a boy in his mid teens, paced in and out of the room, looking to the door every time a dog barked. Siha was right. Everyone was afraid.

Since 1962 Burma has been under military rule. A coup staged by commander-in-chief General Ne Win in March 1962 created a military junta, which has since controlled the nation of Burma. Their tactics are propaganda, fear, violence, and oppression.

In 1988 pro-democracy demonstrations turned into bloodbaths all over the country, when troops were sent to crush the student-activists who agitated for democratic reforms. In the following days, hundreds of unarmed youths died in clashes with the military police, either as a direct result of violence or through drowning or trampling in the chaos. Many dissenters and suspected dissenters fled the country to Thailand, where they called on the international community to recognize their struggle for democracy.

In 1990, under intense pressure from the population, "free" elections were held. Those in exile hoped the results would allow them to return home. The National League for Democracy (NLD) won in a landslide and their leader, Aung San Suu Kyi, was promptly arrested, as were her supporters. Rather than cede power to the elected NLD, the military junta renamed the country Myanmar, renamed their party the State Peace and Development Council (SPDC), and upheld the authority of military rule. Freedom of opinion and freedom of expression are stifled. Dissidents are jailed, tortured, and murdered.

Though Aung San Suu Kyi had not been involved in politics for most of her life, when the opportunity came to restore her father's legacy, to unite the country, which had been under threat of breaking apart along ethnic lines since its founding, she could not refuse. Not only did she become Burma's elected prime minister, a champion of nonviolent resistance, and a Nobel Peace Prize winner, she became Burma's most prominent political prisoner. She has remained under house arrest or in prison for most of the years since she was elected leader of her country in 1990.

Since she has been incarcerated, the military has jailed, tortured, and murdered many members of the NLD, using them as examples to discourage others from joining. The director of information for the NLD, Myint Aung, was arrested in December 2000, along with his assistant. They remain in the notorious Insein prison, where reports of torture during interrogations have leaked out over the years. Aung Tun, an historian of the student resistance movement in Burma, was arrested in 1998 and charged with aiding terrorists. The courts sentenced him to over a decade in prison. In Burma alias Myanmar, peaceful student protests are considered terrorist acts. For this reason, the University of Rangoon is closed more often than it is open.

In May 2003, after one year out of detention, Aung San Suu

Kyi was arrested again, and the military junta resumed a harsh crackdown on democracy activists. As of August 2006, they continue to ignore international pressure demanding Suu Kyi's release. She was forty-five years old when she was elected prime minister. She celebrated her fiftieth and sixtieth birthdays in prison, never having taken office.

In the wild jungles along the Thai-Burma border, a guerilla war is ongoing. The military government blames the violence in remote areas alternately on rebel ethnic minority groups who want the right of self-determination and on supporters of democracy who want to destroy Myanmar's "stability" and "prosperity." But for those who do not recognize the authority of the junta, these rebels and dissidents are freedom fighters. It depends entirely on whether one considers the country a place called Myanmar or a place called Burma and to which ethnic group one belongs, the Burmese majority, or the Karen, Shan, Chin, or Mon minorities.

The State Peace and Development Council (SPDC), which the military junta renamed itself in 1997, continues to clash with ethnic minority armies and political parties fighting for autonomy, such as the Karen National Union (KNU) and the Mon State People's Army, who, in turn, fight with each other for control of lucrative smuggling routes into Thailand.

Cease-fire arrangements with SPDC—the military junta— do little to stop the violence. During a cease-fire with one of the ethnic armies, I was told about military buildup and forced displacement of civilians in that region. Villages are razed and civilians are raped and killed to force their relocation, often to make way for lucrative infrastructure projects like dams and railroads or to cut off the militias from any source of income, support, or recruits. Civilians are forced into government-controlled relocation centers or are displaced internally in Burma. There are no precise figures for internally displaced persons inside Burma, but some estimates put the number around 500,000.

I visited one migrant school near a cease-fire region and worked with all the children at the school on a large drawing of their village. They elected two artists to draw on the board (democracy in practice!) and they told the artists what to include in the picture.

"Houses," they shouted and the artists drew houses.

"Ox," and the artists drew an ox.

"Clothes on the line and flowers."

"No," one of the artists said. "That's not interesting."

"I'm interested in that," I said. "Everything is interesting for me. There are no wrong answers." Everyone laughed that there were no wrong answers. The children usually learned by rote memorization. Even the teachers, who gathered around the edges of the classroom, enjoyed seeing the students shout and get excited describing their village. They allowed the shouting to go on, much to the disbelief of some of the students. The teachers were laughing among themselves, walking back and forth to get a better of view of the kids, of the board, and of me. I was excited, pointing at kids to make sure they got a chance to speak. Some chickens wandered in to investigate, but the teachers chased them out. The artists drew the flowers and the clothes on the line and some chickens.

"Soldiers," one of the girls said. No one disagreed. The teachers grew quiet. The artists drew soldiers. They drew airplanes dropping bombs. They drew bullets coming from the soldiers' guns.

I left the village after some tea and a conversation with four of the students. They didn't talk much about the drawing on the board or the soldiers. They told me a joke that took a long time to tell and, via the translator, wasn't very funny. It went like this:

Monkey was eating a piece of fruit and he told Tortoise to go with him to the riverside to get some more. Monkey told Tortoise to climb the tree and get the fruit, but Tortoise said, "I can't climb trees," so Monkey climbed the tree and ate the inside of the fruit and gave Tortoise the empty husk. Tortoise went off and found his own fruit tree and ate perfumed fruit. Both of them went before the king who liked the smell of Tortoise. Monkey asked where he got that nice smell and Tortoise sent him to a different fruit tree with stinking fruit. When Monkey came back, he smelled terrible and so the king exiled him forever.

We all stared at each other for a while, thinking about the joke they had just told, and then we burst into laughter at how hard it was to tell a joke through a translator. One of the girls said it would be better if we acted it out. One of the boys asked where she would get all that stinking fruit. The translator joked that he was a preacher by profession, not a comedian.

Three days later, I learned that the village was attacked, despite the cease-fire in the area, and that all the children, all the inhabitants of the village had fled into the jungle. I could not find out who was responsible, why fighting had erupted there, or where the children were. I knew only that everyone had run off into the jungle somewhere in Burma. Some weeks later they returned and rebuilt the school, rebuilt the village. This time the school is made of stone and concrete, not so easy to knock down as the wooden stick school they used to have. I wonder if these children know the story of the three little pigs, if they would find it funny or nod knowingly because they live in a world where the big bad wolf often comes knocking.

Land mines litter the regions where members of the minority groups settle, and internecine fighting displaces thousands into the jungle. In addition, the military levies a heavy tax from the people, as a soldier's wages are not adequate. To keep the war going, civilians are conscripted into forced labor, acting as porters or builders for military projects without pay. Many of the porters who are sent with units into fighting are killed. If they cannot carry their loads they are beaten and abandoned. Often, their bodies are left where they lie when the soldiers move on.

According to a recent Human Rights Watch report, the Burmese army recruits more child soldiers than any other army in the world. The report, "My Gun Was as Tall as Me," puts in the number of soldiers under eighteen in the SPDC ranks at around seventy thousand. The children fight, cook, clean, and carry for the army. As economic conditions in the country continue to decay, families are left with fewer options and more children join the armed forces, on one side or another. It can be a method of protecting one's family to have a child in the army, though children are sometimes taken and forced to join as well.

To escape these dangers, many Burmese of dissenting opinions or from minority ethnic groups flee to Thailand. In stable, prosperous Thailand, the Burmese hope to find work, safety, and a future for their children. They find none of these.

Current Thai policy creates arbitrary guidelines for classification of Burmese into various categories, such as "economic migrant," "person of concern," and "temporarily displaced persons." These classifications can be misleading and dangerous for the refugees, as they determine their legal status and level of assistance. Often, the classifications deny the underlying causes of migration to Thailand: civil strife, persecution, and human rights abuses.

The Thai government recognizes some 138,000 "temporarily

displaced" persons, mostly from the Karen and Karenni ethnic groups, whom the Thais have determined were "fleeing fighting," ignoring the nearly 2 million from various ethnic groups that live as registered migrant workers or illegal migrants in the country. Those illegal "migrants" are an embarrassment to the Burmese government, which passed Law 367/120-(b) (1), making it illegal to travel to Thailand without authorization. In the eyes of the Burmese government, all these "migrants" are criminals, and the Thais are quick to agree.

In an attempt to improve relations with their profitable neighbor and avoid a massive refugee influx, Thailand discourages the Burmese from migrating. Given the economic boom in Thailand and the need for unskilled labor, a registration process was initiated for Burmese workers, but the process was expensive and the burden fell largely on the employers, who benefit from illegal and frightened employees. The registration process became a way to crack down on the migrants rather than protect them. Police extortion of the refugees is overlooked, as evidenced by a set of brutal murders in the town of Mae Sot in the summer of 2003.

Migrant children are not given access to school or health care (unless provided by the migrant community themselves), even though these are promised rights under the UN Convention on the Rights of the Child to which Thailand is a signatory.

Siha, his aunt, his mother, his cousin, and my translator are all illegal migrants from Burma, members of the Shan ethnic group. They fled consequences of the civil war and human rights abuses by the government, though they cannot establish that they arrived in Thailand directly "fleeing fighting." They cannot attain refugee status. They have lived without documentation in Thailand since 1998.

"My mom's not afraid anymore, so I am not afraid anymore. The other children's parents are," Siha said. His own mother, who was out of the country at a human rights training at the time we met, I was told, is politically engaged and a confident, fearless woman. Siha seemed proud of his confidence, glad that he had conquered his fears. In conflicts, children often take their cues on how to react from their parents. The level of stress they feel is directly connected to how the adults around them react and can bear little relation to the amount of violence they have actually experienced.

Siha's warning about how fearful other children's parents were was played out in the following days. I tried to arrange visits with families living on construction sites. I saw the level of fear in which the refugees in urban areas lived.

The first person to speak with me was a Thai performance artist who worked with refugee children. "I will do what I can for you," he said. "I can try to help you, but now is a difficult time. The government is cracking down on the migrants. I will talk to people."

He could not arrange anything for me because he was leaving the country the next day to perform abroad. He referred me to a woman who provides assistance to the migrants on construction sites. Call her P——. No real names can be used here, since operations to assist the refugees are under constant scrutiny and might be put at risk. On July 15, 2002, the Thai National Security Council declared martial law on all northern border areas with Burma, banning foreign journalists and NGOs. P——, an affable British woman, is under constant threat of being shut down.

P—— met with me over a soda.

"Now is a difficult time," she said.

"How so?"

"The Thais want to reopen the border with Burma. They make a fortune off the trade. The refugees are an embarrassment to the region, especially because the Burmese suspect many of them of trying to undermine the military authority. Often they are accused of involvement in the Karen National Union or one of the other insurgent armies. Usually they are farmers who could not support themselves in Burma anymore."

The Karen National Union (KNU) is one of the largest ethnic armies in Burma. Since 1949, when disputes over the drawing of the border for the new Karen state led Karen officers in the post-colonial Burmese army to mutiny, the KNU has been fighting for autonomy in areas dominated by the Karen people. The KNU leadership, according Burma scholar Christina Fink, was a mix of university-educated dissidents and experienced soldiers. Boasting thousands of members and income from smuggling and improvised border tolls, the KNU remains a formidable foe to the Burmese government over fifty years after its founding. The KNU enjoys popular support along the Thai-Burma border.

P—— continued: "The work that the migrants do on the construction sites is dangerous and they get injured sometimes, but that is not the worst problem. They have to remain hidden or else they lose their jobs and they can be arrested. They are very afraid of strangers. I do not think I can bring you out to see them. It would draw too much attention."

I asked, naively, if the children could be brought to meet me.

"I don't think so," P—— answered. "These people have seen children taken away for one reason or another and never return. They would not be comfortable with that, and it would scare the children far too much. Even if I could arrange for you to meet them, they might not say anything at all to you. The fear runs very deep."

I tried many more avenues. None worked. While I met with

one local NGO, a field office called the headquarters in a panic. The police were raiding the office. They demanded identity cards from everyone, they searched files and looked for information about any activities supporting the "illegals." Luckily, the Burmese woman talking to the police had Thai identification papers. A wave of relief went through the headquarters. Though these local NGOs operate on shoestring budgets, they must use encryption on all their computer files. The police and the National Security Council threaten to shut down their operations every few months.

Later that day, most of the organizations aiding the Burmese destroyed or encrypted their files. One woman was asked to shut down her operation by the Thai National Security Council because of her involvement with the "terrorist" ethnic groups in Burma. She also learned on the same day that there was a price on her head. Someone did not like what she was saying about the military junta. The SPDC sees aid to the refugees as aid to their enemies, and this threatens trade with Thailand, which cannot be seen as harboring its neighbor's enemies. Powerful people want the refugees to vanish.

The children on the construction sites *are* vanishing. They live in hiding. With the few exceptions Siha mentioned, they do not go to school. They have little contact with the outside world.

In Bangkok, I had more luck talking to some of these hidden children. An NGO agreed to take me to see some families in their homes, provided the families were willing. We would have to travel very low profile. Of medium height, white, and not speaking a word of Thai, I was not sure how I would maintain a low profile in the slums of Bangkok, but early on a Tuesday morning I found myself on a narrow street beneath a massive concrete building. A variety of smells overwhelmed me: sandalwood, sweat, excrement, exhaust fumes.

We walked through a shop of some kind. Two men and an older woman followed me with their eyes as I passed, bowing my head respectfully. A blind dog rested by the door. An old television blared. A shrine to the Buddha sat next to the television, its face partially obscured by the rabbit ear antenna. As in most Thai businesses, a photo of the king hung on the wall. They turned back to their television as I walked out the back door without a word.

We emerged onto another street below an even more ominous concrete building. A whole gang of mangy dogs waited by the front door. Their skin hung off their faces; their ribs were visible and their fur was matted and bald in patches. One of them looked up at me with bloodshot eyes, raising its muzzle from the dirt and growling as I passed the entrance to their building. The way their thin limbs twisted and splayed together made me think of Cerberus guarding the gates of Hell. This building was not our destination either, and the dogs did not bother to follow us. We slipped into a narrow alley. There was a small Buddhist shrine at the entrance to the alley, decorated with flowers and burned-out candles. Just beyond it, two red-faced Americans in dark blue jackets were speaking to a man about the Book of Mormon. They must have sweltered under their blazers. I nodded at them, acting as if we knew each other and creating a credible alibi if someone became suspicious of my presence.

Through the alley was another building, more decrepit than the two before it. This was our destination. Again, sickly dogs acted as sentinels. A woman standing in the shade of the doorway came over to us. My guide and translator, himself an illegal refugee, spoke to her very quickly. She joined her palms together in front of her lips in greeting, and I did the same. Then she shuffled us inside and up to the twelfth floor. It was very quiet.

The last door on our left opened to an unfurnished room

lit by buzzing fluorescent lights. We removed our shoes in the hallway and entered. There were mats on the floor to sit on, and I could see a small kitchen and bathroom. Compared to the privations of Africa, this urban refugee setting seemed almost desirable. Five children assembled in front of me with their parents. Their father's right leg was made of plastic. Very quickly, soda and crackers were brought in. The family welcomed me gracefully. No one seemed too fearful. Then the door was shut firmly and locked from the inside. My translator introduced me, explained that I would like to speak with the children about their lives and their perceptions of refugee life, and have them draw some pictures. The father and mother joined their palms in front of their lips, and I did the same in return. Then the children and I returned the gesture. The kids giggled hysterically, and the father and mother shot them a glance. It seems I had joined my hands together at the bridge of my nose, a sign of respect reserved for monks. The children thought it was one of the funniest things they had ever seen. When I smiled at the explanation from the translator, the parents relaxed. They would not want me to be offended.

The children were eager to draw and the parents were eager to talk. The family had not left the building and had hardly left the room in several months, not for school or work or play. I rapidly learned of the slow torture that this life entails. While these kids are not in the combat zones of Congolese child soldiers or even the children internally displaced in Burma, they have their own battle to fight, against depression, silence, and disappearance.

"A few days ago our neighbor threatened my oldest daughter," the father said. "She was singing too loudly, and the neighbor threatened to call the police and send us back to Burma or to the border area. This cannot happen, of course."

The children remained quiet while their father spoke, concentrating on their drawings, signaling each other with looks to exchange colors.

"We are at the mercy of our neighbors because we are not recognized by the government. We registered with UNHCR (United Nations High Commissioner for Refugees), but they said that the government was not recognizing any more refugees at this time. I am not legally allowed to work, but that matters little. . . ." He gestured at his leg, shaking his head. "None of us speak the Thai language. The children cannot learn it. They cannot go to school." He rubbed his eyes with weariness. "They do not leave the house. It is too dangerous. If they play too loudly and annoy someone, we can be arrested."

I felt lost here in a world without play. There was a chasm of experience between these children and me. I wanted to leap it with a game, if not soccer something quiet, something mindless and fun that we could do together, but no one felt much like playing.

My translator, who drives around the city for one of the aid organizations delivering monetary assistance to the illegal Burmese refugees, leaned over and told me that this family would have to be moved.

"People complain that they have too many children and they cannot pay their electricity bill. We give them some money, but it is not enough. We have to be careful, because our assistance program is also illegal. No one here wants these people to be helped. No one wants them to stay."

"Do you know where you will go next?" I asked.

"We will go where we find a place. We do not know where," the father said. He leaned back on his hands, stretching his fake leg and rubbing the area above it.

"Why did you leave the border area? Is it not safer there?"

"I fought the Burmese government with the KNU. I disagreed with my commander and he tried to kill me. I was shot in the leg. I cannot go back to Burma because I am their enemy. I fought for democracy. I cannot go to the border area because of the KNU. I must seek shelter in Bangkok, even though we must live like this." He gestured at the room around him. In the half hour I had been with this family so far, I felt the light squeezing my eyeballs more and more. The walls inched in around me from minute to minute. They were bare and white. This was a prison cell.

"I want to study," said Thinzanoo, the oldest daughter, the singer whose song had put the family in a precarious situation. "Now I help my mother all day and raise my little brothers. I don't see any other people. I want to have a good education."

Education, schooling, the bane of so many children in the "developed" world, is Thinzanoo's hope, as it is the hope of so many of the children I met in Africa who had been uprooted by war. School gave children companionship and a space to act like children. It represented their best hopes for the future and an alternative to their present suffering, exposed to the adult hardships around them.

A few days later, another girl Thinzanoo's age drew a picture that illustrated the role school plays in the inner lives of these dispossessed children. May eagerly showed me her skills in English when she drew a picture of a girl named Susu and labeled it: "I am going to school." Then she drew a picture of another girl, her hair done up, in nice clothes with a professional air about her, happy and pretty. This drawing said: "I went to school." May looked like the first girl, Susu, but dreamed of the future in which she would evolve into the other girl in the picture, the one who had gone to school (Figure 8).

The kids in hiding want stability and opportunity. School

shows them a way into a good job, into security and comfort, and into the society that they see and hear other children entering every day and that they are forbidden to enter. When the days stretch out without any change, without any hope, the simple wish for education can become an escape route, a route that their parents can rarely provide.

Thinzanoo was ashamed of how her family lives. When I asked about the war in Burma, she gave no answer, and tears formed in her eyes. When I asked what she likes to do, what she wants to do when she is older, the tears burst.

"She weeps," my translator whispered, "because she is ashamed. She cannot think of an answer." Her attitude echoed the somber mood of her father, and I thought of Siha, how he took his cues from his mother, and I looked at Thinzanoo's father. Her younger brother, Ostar, drew a picture of their father (Figure 9). In the picture, he has short, neat hair. He wears a tie and a clean pink shirt. The man sitting beside us looked nothing like this. His prosthetic leg was dirty and chipped a bit. His hair was long and tangled. He had heavy, dark bags around his eyes. The children, very hesitant to speak with an outsider, deferred everything to their father. They looked on him with respect and admiration. In his face I saw weariness. He was not the man in the picture anymore. Fleeing his homeland, he had left that man behind. Now he found himself helpless to protect his family, depressed and uncertain. His own fear of the border, of the outside world infected his children and made them extremely nervous as well. I wondered how greatly their outlook would improve if their father could receive counseling or an opportunity to restore some of the confidence he lost when he lost his leg.

The children have no future. They cannot go home, they cannot settle in Thailand. Their entire world has shrunk to the size of the room on the twelfth floor of a Bangkok tenement, their worldview shaped by their melancholy and anxious father. They do not fear the mangy dogs outside or the crime in the bustling metropolis because they do not know it; they never experience it. They never experience anything outside.

I asked them to draw pictures of Bangkok as they saw it. The children's drawings showed no resemblance to the city itself.

"I drew a mountain and the sun rising over clouds. I drew our house and a tree with a woman. The woman is growing a flower," said Thinzanoo. She drew rural Burma, essentially, a pleasant scene from a life she desires, a life that has been closed off to her (Figure 10). Without the ability to assimilate into the new country, her only resource for images in her imagination is the memory of the past.

Their father looked at the drawing and smiled, his own eyes growing moist. His children were being erased from the earth, disappearing from reality in their fluorescent little room, and he was helpless to stop it happening. He looked at the drawing of Burma for a long time without speaking. The whole family looked quietly at the serene pictures of home, drawn from imaginations battered and bruised by loneliness. The whole family was homesick for the land in their imaginations, homesick for the land of pictures.

Another family I met later that day found themselves in an even worse situation. They were not registered with UNHCR and were running out of time. M—— and his wife worked for a pro-democracy organization in Burma, fighting for freedom of expression and the people's right to self-determination. They are members of the Burmese ethnic group, the majority group in Burma. When economic conditions became too harsh and the threat to his family due to his political activities became too great, M—— fled Burma and arrived with his wife and two children in Bangkok. They lived in a windowless eight-by-ten room, also under the harsh glow of fluorescent lights. The neighbor's television blared through their walls. It sounded like a violent action movie was on. The family had lived in Thailand for five months.

"I can't play too loud or the police will come," M——'s ten-year-old son, Caleb, said. M—— explained that just the day before, one of his neighbors was arrested and sent back to Burma with his family because the Thai neighbors complained to the police. The children were playing too loudly, shouting during a hiding game, which annoyed the older people on the floor who just wanted to watch television in peace.

"I do not know what will happen to them now," M—— said.

"I'm afraid of the police and the Thais," Caleb said as he played with the toys his parents brought with them. "I want to go to school." Even if it was legal for him to attend a Thai school, even if he spoke Thai, he could be arrested if he ventured out onto the street, or even lingered in the hallway of his building for too long. An aid agency that provides help to these families told me that of the 160 children under their care in Bangkok, they have successfully managed, with the help of UNHCR, to get two enrolled in school, with no guarantees for the children's security from the police.

Caleb's only forays into the outside world involved going across the street to the market with his mother. He did this about four times a week. The trips lasted about forty-five minutes. They were his only time in the sunlight. His skin had grown sallow.

"We leave the light on all the time because there are no windows and it is so dark in here. There is no ventilation. I worry about the health of my sons. We have moved four times already, to avoid the authorities," M—— said. "The children do not feel safe, as they know we will move again very soon. We are always moving."

M—— cannot work in Thailand so his family relies on the assistance of NGO's illegal operations to help them. These programs can be shut down at any moment. Their offices, like all the NGO offices, are watched. The families receive cash assistance for food and health care, but there are no programs to help the children, no opportunities for the illegal children in Bangkok to socialize and interact with other children. They live in fear and in hiding. Now, after five months, M—— can no longer receive assistance as an unregistered migrant. UNHCR, at least, must recognize him so his family can continue to receive aid. For now, he has to try and find work, where he will most likely be exploited. If he is not paid, if the conditions are unsafe, he has no recourse with the law. There is nowhere he can turn.

For these children there is little hope for the future. Their parents fear returning to Burma, yet Thailand will not accept them. Of the twenty children I interviewed in Bangkok, not one felt safe in their new surroundings and every one wanted only to go to school and to play with other children. They have no homeland, no community. Their parents are nearly helpless to support them. The outlook for their future extends only as far as the four walls of the rooms in which they try to survive. Their drawings all look the same: dream worlds of home, as it probably never was, with the peaceful hills of Burma gently fading into the sky (Figure 11).

Outside of the cities, in the areas near the Thai-Burma border, the problems faced by the children were similar but the mechanisms for survival changed, and support from a larger community of migrants strengthened them. In border areas, entire economies have developed around the refugees, economies that support the police, a variety of smugglers, border guards, and gangsters.

The border town of Mae Sot is a rough and tumble place. According to the *Lonely Planet* guidebook, there used to be a billboard in the center of town that read: "Have fun, but if you carry a gun, you go to jail." The illegal gem trade flourishes. In a café where I liked to spend my afternoons sipping beer and pretending to be a tourist, I watched many packages of gold and jewels change hands. Men were always coming and going, drinking a beer or eating quickly while they juggled cell phone calls and exchanged glances at the street.

During the day, the town was charming. Twenty-foot-tall golden statues of the Buddha rose from the jungle, their heads jutting out above the tree line and glistening with beads of water. It rained every day in September. The white minarets of the Masjit noor-ul-Islam mosque rise behind the low buildings. Every day I would hear the muezzin issuing the call to prayers, which echoed through the streets, through the trees, and drifted to the hills in the distance.

Allah-u-Akbar Allah-u-Akbar! Hayya 'alas salah Hayya 'alas salah.

The municipal market sat behind one of the three mosques in the city. On my way into a narrow alley that opened into the market, a group of street kids surrounded me, poking and pulling at my clothes. I love markets because they tell you secrets about a new city, and they are always filled with eager children. This market was no exception.

Their faces were smeared with sandalwood powder, a traditional Burmese practice. The powder has many functions: it's good for the skin, it keeps one cool, and it's a fashionable thing to wear. It also expresses certain cultural affiliations. In Bangkok, the Burmese could not have worn it on the street, even if they spoke Thai. It would have been a dead giveaway and gotten them thrown right into jail or deported. In Mae Sot,

I noticed the Burmese walked openly on the streets, speaking their language without fear. I gave the kids a little candy that I kept in my pocket, made a few faces that they didn't think were very funny, and turned into the market. The kids were having a great time laughing at me as I walked away, though they quickly returned to the main street where they had more room to kick around the ball that one of them held under his arm.

In the market the smells were overwhelming. The tables overflowed with roasting meats, raw fish, fried fish, salted fish, raw chickens, live chickens squawking and pooping in baskets. Women shoved roti and chapati in my face, ladles full of curry, buckets of live eels. They offered T-shirts and cassettes and sandals. The street burst with shops, tables, stands, and carts. People squeezed between them, bikes dodged through the crowd. Everyone cleared the way for a passing police motorcycle, keeping his or her head down until the bike got through.

"Security service," a boy selling T-shirts said to me in English, making a little gun with his hand and laughing. "Boom-boom," he said for some reason. His T-shirts intrigued me. They bore hagiographic images of Thai and American pop stars and beatified pictures of Osama bin Laden. The boy behind the table wore one of the Bin Laden T-shirts, perhaps oblivious to its meaning to me. He made a peace sign. It happened to be September 11, 2002, exactly one year since the terrorist attacks. When I walked away, the boy shouted after me, still in English, "Have a nice day!"

Everyone I passed gave me a once-over. There were a few beggars, but mostly the shops bustled with loud commerce and peals of laughter. Over half the people I saw wore face powder. Discussions cascaded over each other. After Bangkok, I was shocked to see so many people living unafraid. It was not until nightfall when I saw the security services again that I understood the way the town really worked.

Two colleagues and I were crowded onto the back of a little motorbike. I was the most junior of the group, so I hung precariously off the back. We set out to get a drink at a local bar popular among expatriates. The roads in Mae Sot all seemed to be one way. I marveled that the same malicious urban planner that made driving in Boston impossible had found his way out to this Southeast Asian border post. In order to get where we wanted to be, we had to go down the main road past our destination, turn onto a small side street, and cross to the other main road that went the opposite direction. The side streets were unlit, and our hearts froze for a moment when we saw a group of uniformed policemen standing in the middle of the small bridge that we had to cross. My mind raced back to the checkpoints of the Congo and unscrupulous soldiers with submachine guns. Checkpoints can be the most dangerous places in the world. In the Congo, a few wrong words at a checkpoint had nearly landed me in jail.

I clung to the back of the motorcycle and held my breath, each bump threatening to knock me off. We slowed, but did not stop as we passed. The officers looked us over but made no move to halt us. Thailand relies on tourism. We continued on.

We missed the bar and had to double back again. None of us were happy about crossing paths with those policemen again. I held on tight and we approached the bridge. That was when I saw what the policemen were doing. They had stopped a Burmese man walking at night with a sack over his shoulder. For a Burmese man, coming from a land of military oppression and vanishings, to encounter a group of uniformed men on a dark bridge is a harrowing experience. As we passed, the soldiers circled the man, blocking him from our view. They were shaking him down.

"The police," my colleague told me, "request to get stationed here. They're all getting rich." If the Burmese can't pay up, they can be arrested. The family must come up with money to pay off the police and retrieve their relative. Otherwise, the unlucky migrant will find himself or herself deported, or worse. We learned from a local NGO that the slave trade is flourishing, with over a hundred Burmese sold into slavery every single day. Children were certainly not exempted from this and were all forbidden by their families to go out at night. Every Burmese child I met said that they did not go out a night for fear of criminals *and* fear of the police.

The next morning I went to a migrant school outside of town. The school sat in the middle of a field. My guides were a group of students who arranged human rights education and democracy training for the exiled Burmese.

It was crowded and stuffy inside the one-room school building, which was made from leaves and bamboo. There were too many students in too little space. The headmaster and his two teachers greeted me at the door. They offered me a seat outside in the shade of the building. The sun shone, and we thought it would be good to take advantage before the rain came. So my translator and I sat and waited for some students to be brought to us. Through the openings in the bamboo, I saw children's curious fingers and wide eyes peering out to get a glimpse of me. Most of these children's families had been subsistence farmers in Burma, and they had little exposure to the world beyond the distant hills. There was a lot of excitement in the air, especially when I brought out markers and paper to draw on. The school had few supplies and everyone, even the teachers, was eager to make use of the new materials. A group of children were sent out. Before anything could begin, I distributed paper all around.

Eleven-year-old Nicholas had wide almond eyes and wore soccer shorts and a frayed jersey. We looked at his drawing—the frightful crucifixion of a villager, bodies falling from the sky, a little boy in purple hiding behind a tree. Nicholas told me about the SPDC attack on his village and his recurring nightmares in Thailand. He had trouble sleeping because he was afraid. His mother tried to comfort him when he couldn't sleep, he said. She told him he was safe now, that it was okay.

Her assurances were not entirely true.

Nicholas said he still had bad dreams, "but not for a while when my mom comforts me." It was hard to ignore the tension in the air, and the children, like the adults, seemed very alert to it. Every child I met mentioned how it was not safe to go out in the town on their own, not safe to go out at night. I imagine the strain can be very tiring for a child and that the bad dreams will not go away until Nicholas is secure, his future stable.

He said he wants to be a teacher when he grows up. He liked his teacher, who told the children about their history and their language.

Nicholas is a member of the Karen ethnic group, which shows little sign of ending its fight against the SPDC in Burma. The Karen have been fighting for an autonomous region, off and on, since the 1950s. The military cracks down on the Karen areas, on the civilian populations, as part of what is called the Four Cuts Policy. The policy aims to cut rebel factions off from food, money, intelligence, and recruits, the four factors that allow them to continue their war against the military junta. When the junta suspects rebel activity in an area, they will enter and forcibly remove the civilians. Anyone who stays behind is considered an insurgent and will be shot. Karen villages are often burnt to the ground, heavily mined, and the inhabitants moved into resettlement areas that, by some accounts, resemble concentration camps more than villages. Some of the resettlement areas are better than others, and sometimes, even after the military allows them to return home, people stay in the resettlement areas because they have nowhere else to go. People who do not go into the resettlement areas find themselves internally displaced in the jungle, with little or no access to aid, vulnerable to disease, starvation, and forced recruitment into either the government army or the rebel army.

Without a population to draw on for support, the junta hopes that the rebel armies will wither. The same idea was used against the Kurds in Iraq, the Albanians in Kosovo, the Dinka in southern Sudan, the Tutsi in Rwanda, and the tribes in Darfur. Kill all the people to eliminate the rebels that may or may not be among them.

The education system in Burma attempts to eliminate notions of ethnic identity from the young and to create loyal cadres for the central government. Ethnic minorities are forbidden to study their own language in state schools. Though approved in 1967, Karen language textbooks were not printed until 1980 and, even then, no new teachers were hired to teach the language. In classes, none of the history or culture of the minority groups is taught. The junta hopes to assimilate the minority groups into the majority Burmese culture, eliminating claims for regional autonomy. As with the political and social spheres, the junta wants to control cultural life of the people. The educational program of the central government serves as a kind of ethnic cleansing, aimed at erasing future generations of ethnic nationalists.

Nicholas's ethnicity placed him in danger and, despite any ideas he may have formed on his own, aligned him with a particular side in the conflict.

I met a ten-year-old girl whose family had taught her to speak

Burmese rather than Karen so that she could pretend to be a member of the Burmese majority, just as Melanie had learned to speak Swahili rather than Kinyarwandan so she could pass as an ethnic Hutu.

"I came here to stay with my grandmother and attend the school," the little girl said. "My brothers and sisters cannot go to school in Burma. We have no money." While she framed her reasons for leaving in economic terms, the causes of those conditions were the results of conflict and persecution. She is Karen and therefore vulnerable to the Four Cuts, to attack and forced labor, and to a hostile education system. Rather than await the destruction of their village, forcing them into resettlement areas or into the jungle to starve as internally displaced persons, her family took her to Thailand.

In Thailand, regardless of ethnicity, it is not safe to be Burmese. Many of the children at the school near the border were born in the refugee camps on the border. Kyaw Win, another eleven-year-old boy, was one of those children. The Burmese army attacked his village, and his family sought safety in Thailand. They were placed in a refugee camp from which the KNU was rumored to operate. The camps were fertile ground for new recruits. In the schools, the teachers could expound Karen history or political thought. Ethnic nationalism developed easily, though the KNU has abandoned calls for independence and now seeks only regional autonomy in Burma. Sitting near the border, the camps acted as a convenient base of operations for attacks against Burma. Threatened by these camps, the Burmese military burnt them to the ground, forcing the inhabitants to flee into the Thai countryside, where they were often unwelcome. Thailand fears the buildup a huge refugee industry and does not want to make any of the refugees terribly comfortable.

Residents of the camps have little access to health care.

An American girl I met who was teaching English in one of these camps told me the story of a student she had recently lost. He grew ill very suddenly. There were no medical facilities in the camp. It was an arduous journey through dangerous jungle roads to get to the hospital. He was not legally allowed to leave the camp, and permission took time. When he eventually did get to the hospital, it was too late. The boy died at age sixteen. When I spoke with his teacher, the family was fighting with the hospital to release his body. The hospital wanted several thousand baht, hundreds of dollars, to return his body to the family. The family, living off a meager amount of aid, and stuck in the camp, could not afford it. They were raising money, but having trouble getting enough. Even the dead Burmese had no rights in Thailand.

A spy sat outside the lobby of the hotel where I was stay-ing. He was watching the activities of foreigners. He wore a white dress shirt, slacks, and dark sunglasses. In the morning he would watch the guests coming and going. I had been warned that there were many people watching in Mae Sot, that it was a good idea to be discreet.

My guide was late one morning and I sat nearby writing in a small journal. The man in dark glasses came to me and looked over my shoulder. I closed the book and walked away. That afternoon, when I returned from my interviews, he took a great interest in my day, asking many questions in decent English. I told him I had been sightseeing, that I was backpacking through Thailand, but he continued to pry.

"Where did you go?" he asked.

"I explored the market."

"Careful," he said. "There is much crime in the market. You

should not go without an escort. I can go with you. What do you do in Mae Sot?"

"I'll be all right," I said, ignoring his question and excusing myself. The man was still lingering the next morning. He watched my guide very closely when she picked me up. Foreigners had been banned from the refugee camps a few hours away, and all arrangements to go in had to be kept extremely quiet. The man in dark glasses wanted, I assume, to make sure I was not violating the ban or meeting with political groups whose activities caused trouble for the Thais. Every child I spoke with was put at risk if I was followed. I would not come to harm, except maybe by losing my visa, but the risks for the Burmese in speaking to me were very great.

Back at the offices of the youth organization that had taken me to the school, we spoke with the doors and windows closed, despite the stifling tropical heat. They had "officially" suspended activities for the time being because of the police crackdown and were trying to present the appearance that the office was closed. A poster proclaiming the rights of all people to self-determination hung on the wall above my head.

"The Thais do not want our programs to operate," my translator said. "We teach about democracy and women's rights. This is seen by the Burmese government as insurrection and, therefore, Thailand is seen as harboring rebels. This strains their relations."

Categorizing most Burmese as economic migrants denies many of the rights to which children are entitled by international refugee law, such as education and healthcare equal to the standards of the host country. While on the surface it seemed that many families had come to Thailand to find work, protracted discussion with children who had arrived recently revealed the danger of life caught up in the Burmese conflict:

Nicholas's drawing of the attack on his village, Siha's drawing of the dangerous border crossing, the families who have chosen to live in isolation in Bangkok rather than face the dangers of Burma or the border areas.

One look at their drawings showed the deep impact the political turmoil has had on their lives. In one drawing by Win, an eleven-year-old boy who also suffered nightmares of the violence in Burma that he had witnessed, a detailed representation of a machine gun points at the flag for the Karen National Union (Figure 12). Violence and flags were linked in his imagination.

Many young people expressed their allegiances through the flags they represented in their drawings. These children were not neutral in the conflict. When asked why war tears at their homeland, Win said simply: "The Karen are fighting for freedom from the Burmese."

Many other children added their voice of agreement to his. They agreed with the cause of the KNU. Few children I met said they knew the reasons for the war, but all who are members of a minority ethnic group knew what side they were on. Kin Wa, a fourteen-year-old girl who wanted nothing more than to go back to Burma—"Thailand is not my country. I was born and raised in Burma and want to return"—wanted to work in an office when she grew up. Not just any office. She wanted to be like her father and work for the Democracy Movement for Burma, his job, of course, being the reason he'd left for Thailand in the first place. She admired her father and wanted to follow in his footsteps.

"If Burma becomes a democracy, we can go back," she said, repeating, I imagine, what she often heard at home. For the world in which she lives, family affiliation or ethnicity is enough for her to have a side in the conflict. Her family life and her political views are tied together, and politics plays a role in her worldview as much as any other lesson she learns from her parents, such as respect for elders or how to do her chores. Politics is not separate from her childhood; it is part of her childhood.

In her drawing, underneath a gentle sun and blossoming trees, she drew a monkey hanging from branches, children playing soccer, and a bright and colorful schoolhouse (Figure 13). The flag of the KNU flies next to the schoolhouse. In black and white, two houses burn, a third house, still in color, is engulfed by monochrome flames. Soldiers with machine guns fire on the village—she chose to use pencil and not markers to represent the violence disturbing the colorful lives of her people. Even the monkey is not safe from the soldiers Kin Wa added to her drawing. Soldiers pump the monkey full of bullets.

"The monkey dies too," Kin Wa said. Along the road there are black and white stick figures, none of them wearing the hats she drew on her soldiers. "Civilians," she said. She did not say whether or not they were dead along the road or if she had just left them uncolored, but it looked to me as if all that was lifeless or destroyed was depicted in black and white. By putting the civilians in color, the KNU flag in color, and the SPDC soldiers in black and white, Kin Wa gave clear representation of her political point of view.

Given their indiscriminate targeting of villages, if you are not on the side of the junta you are on the other side, and the exiled children of Burma could not help but take a side. Some of these children will grow up to join one of the revolutionary forces, while others, like Kin Wa, might become political activists. Many of them admired Johnny and Luther Htoo, the God's Army twins, and considered them national heroes.

But there are others who feel no such connection to a cause, despite the persecution they have suffered.

Aung Su was ten years old. He had lived in Thailand for one year.

"I used to live in Mon State, but I came here with my mother one year ago. She wanted to work here and wanted me to work. It's very difficult to earn money in Burma."

Judging by this first answer, Aung Su's family were economic migrants, unregistered and illegal, without any protections afforded those "fleeing fighting." He did not mention political reasons for fleeing, though they perhaps existed. His disengagement with a nationalist cause might come from his mother's disengagement or the fact that some parents choose not to discuss politics with their children.

"It was hard to cross the border. There are robbers, Mon soldiers"—the Mon are another ethnic group in Burma—"Burmese soldiers, Karen soldiers, all wanting money. We had to pay many times at many checkpoints. It was dangerous."

The gamble of encountering hostile or unscrupulous soldiers while traveling through conflict ridden areas away from the safety of one's own village seems like a big risk to take with a ten-year-old, unless the risks of staying put are greater.

After about half an hour talking together about sports and movies, school and dogs (he hates them, because they chase and bark; I told him that I miss my childhood dog, which he found absurd), Aung Su began to describe life in Burma.

"There are lots of problems with soldiers. They search for young men to take as porters. Many troubles. We came here. They came to my village and took young men and put them on a truck. They took rice too. My uncle climbed a tree to get away from the soldiers, but he fell and died. A ghost pushed him out. Another uncle died in battle as a porter. They don't get guns to protect themselves. Many die."

This description of his situation in Burma presented another side of the story. The reasons his mother could not earn money in Burma were directly connected to the fighting. The threat of theft by the military and of forced conscription loom over everyday life. Aung Su has a bright smile and enjoys talking. He is more outgoing than many of the children I have met. He works at a small shop in town when he is not in school. He recently saw the Spider-Man movie on his employer's television set and drew me several pictures of the red-and-blue-clad superhero.

"I like to draw Spider-Man, but I don't usually have colors. I like this." He held up one of the pictures. "It looks more like it should with color. I am happy when I'm drawing" (Figure 14).

"What would you do if you were Spider-Man?" I asked him.

"First I would go to the Thai gangs. They attack people at night. They rape people and rob them. I would get rid of those gangs."

"Would you go back to Burma?"

"I don't want to go back to Burma. I want to be an English soldier. I want to drive and have lots of guns."

An English soldier? He said he could not explain why. He just liked English soldiers, by which I think he meant American soldiers. (He thought I was English, so perhaps he was telling me what he thought I wanted to hear.)

Aung Su is like most boys anywhere. When I was ten, I played with G.I. Joe. I liked the idea of guns and soldiering, driving big cars and blowing things up. But Aung Su has seen war up close. He knows that it kills and destroys villages. Knowing this, his young boy visions of guns and cars are still intact. Is this the power of the media, which still reaches him through his boss's television? Is this an innate quality in young boys who want to be bigger and tougher than they are? Is Aung Su angry at his powerlessness to save his uncle from the army that chased him into the tree, from the ghost that pushed him out? Is it power Aung Su wants, and if so, isn't it power that those soldiers who came to his village wanted too? After all, he who has a gun does not go hungry.

His animosity towards Burma, towards going back at all—he wanted to be an *English* soldier—is itself a statement of his political loyalties. He wants to fight. He does not want to fight for *them*, neither the government that took his uncle nor the rebels who scared him and his mother on their journey. Spider-Man is a lone hero, which seemed to be another part of the fantasy Aung Su liked to entertain. The Thai gangs, which are known to harass the Burmese migrants, were Aung Su's main concern. He had put Burma behind him, adjusted to the new environment, and adjusted his concerns to the problems in the new environment. The Myanmar army and the regime, for him, were a distant abstraction.

Children, however, are not an abstraction to the regime. Their education program acknowledges the role children can play in the future of their country, as dissidents or as loyal cadres. The army employs an estimated thirty thousand child soldiers, the highest number of any one army in the world. The junta has an acute awareness of children's capabilities.

The young are just as capable of dissent as their parents, as Kin Wa and Win showed me, and are just as dangerous to the military government. Aung Su, even though he didn't seem very concerned about the situation in Burma, had forged aspects of his personality in reaction to the conflict. Even though he wanted to be English, he could not escape his identity, and the policies of both Burma and Thailand will not let him or his peers define themselves any other way.

Sum Chai, a friend of Aung Su's, lived his whole life in Thailand. His parents used to be plantation workers in Burma but now they worked in Bangkok.

"I'm worried for my parents' safety in Bangkok," he said. "There are a lot of drug addicts in the city. I live with my aunt now. I worked on the construction sites with my father in Bangkok, carrying bricks. We made enough money to get by, I think. My mother would not let me stay with her because there is no school for Burmese children in Bangkok. I want to go to school and be a doctor in Thailand."

"Why not in Burma?"

"I've never been to Burma. I've never seen a hospital in Burma. I want to stay in Thailand. There is more to see. My parents want to go back to Burma, when it is safe, but I like it here. I can speak Thai and Mon and Burmese. I would also like to study English."

Sum Chai echoed a lot of the Burmese children. As far as national policy is concerned, he is Burmese. He lives with other Burmese. His family is Burmese. But he has grown up in Thailand. This is his home. He was not angry at Burma or at Thailand. Unlike his peers, he said he's never been afraid of the police. (I don't imagine he would have admitted being afraid of anything, not in front of his buddy Aung Su.) His predicament is simply that he has no nation. He is unable to hold property or receive a certificate of higher education. He will grow up trapped between the policies of a military regime and the policies of his country of refuge.

Together, they will deny him safety, schooling, employment, culture, language, and history. He is defined by all of these outside forces and labeled by them. He likes soccer and wants to learn English so he can study computers. His friends want to go to good schools and walk home unafraid, to become doctors and teachers, to work for democracy, to travel, to be superheroes.

When we spoke in the muddy schoolyard out of sight from the prying eyes of the police on the road, he was ten years old and already knew none of them could do any of these things.

FOUR

"I Am Getting Used to Living Here"

Children in Camps, Shelters, and on the Streets

It was nine a.m. in Kakuma Refugee Camp and Charity was not in school.

On the way into the camp near the Kenyan border with Sudan, there were six boys squatting in the dust around a rough wooden toy, a handmade car. They fiddled with the pieces of it, hewn branches and scraps, trying to make them fit together so it would roll correctly. Up the road from where they played, men unloaded trucks. Each of these trucks contained several tons of food, meant for distribution to the roughly 80,000 refugees from six nations who lived in this camp. Dust covered everything, creeping into the folds of clothing, inching up the walls of buildings, seeping into the cracks of dry skin on my hands and face.

The sun had already started its brutal arc into the sky and shade was scarce. It was the rainy season and so allegedly cooler, but it was above ninety-five degrees Fahrenheit. It was above ninety-five degrees Fahrenheit the day before. It would be again tomorrow. Because of the rains, there was less dust than usual. At certain times of year, I am told, the dust blots out the sky. There was some vegetation on the ground: scattered patches of what looked like dandelions bursting from the dirt, the first blossoms of what would become brambles of two-inch thorns.

A small cress-type leaf that starving Sudanese youths and Somali livestock eat has found its way into the crowded graveyard at the entrance to the camp, near where the boys are building their car. Sometimes violence erupts over who gets to eat the vegetation first, the goats or the Sudanese children. A few acacia trees stand in defiance of the hot and cracking ground, each by itself with no clear relationship to the others. Each tree is precious, because shade makes the best place to sit. Light, heat, and breeze dictate the real estate market in the desert. Leave the shade of one tree and who knows when you will encounter the next. The landscape is merciless. There isn't enough water. Or rather, there is just enough for the few gnarled trees and no more. Each tree burns alone in the sun, defying the most human of desires: to gather together.

My translator, Simon, and I walked down a dirt path coming from the schoolhouse where we had hoped to speak with Charity. Simon is one of the famous Lost Boys of Sudan, who traveled on foot to Ethiopia and then from Ethiopia back to Sudan and then to Kenya, suffering bombings, starvation, and crocodile attacks. He was goaded forward by rebel soldiers who wanted to control the food aid coming to the parentless youths and by the bombs from the government in Khartoum that continues its campaign against the black Africans as I write this, now focusing on the Darfur region.

Simon has lived in the Kakuma Refugee Camp in Kenya for the past ten years, since he was fifteen. Boys are especially vulnerable because they are considered potential soldiers and because they tend the herds of cattle, and to destroy them is to destroy the southern Sudanese way of life. Simon is about six and half feet tall and walks with the loping gate of one who is used to walking everywhere. He has a warm smile, which does not betray the horrors he has seen in his twenty-five years of life.

We walked quickly, pouring sweat, and arrived at a gate made from oil tins donated by the World Food Program. Someone had pounded the tins flat and stuck them together to make a door. The gate was surrounded by thorn bushes to mark off the compound. Stagnant pools of water shimmered in the sun. We slipped through the door into the compound of low mud houses clustered together. No vegetation grew here. A group of small children leaned against one of the houses talking excitedly. When they saw me, one of them shouted: *"Khawaja! Khawaja!"* This is the Dinka word for white person. In the Tanzanian camps, the word was *mzungu,* and I had gotten used to hearing it chanted by flocks of children with bright and embarrassed smiles on their faces. The other young children joined in the *khawaja* chant and I waved, which resulted in an explosion of laughter. An older woman came out to see what the commotion was about.

"Hello, Mama," I said, the "Mama" a term of respect. She nodded, but appeared skeptical. The usual conversation that I had with countless strangers throughout East Africa about her family and mine, our health on this day, whether or not it was my first time here, did not occur. It was too hot, I suppose. She just looked Simon and me up and down.

I tried to make an introduction, but she showed no interest.

"We are looking for Charity Anyieth," I said. "I was to meet her at the school, but she was not there."

The woman did not respond, but turned over her shoulder to one of the small boys, now silent but fixing an unblinking stare at the *khawaja*.

"Charity," she said, and the boy ran off to one of the little houses. A moment later, a tall and elegant young woman emerged. Charity. She saw me, though she did not smile. She came over and shook my hand.

"Charity, my name is Charles. I'd like to talk to you if you have a chance."

"Yes," she said in English. "I have been told you will come. I will meet you up at the school. I have to finish my work."

I agreed and was promptly ushered back through the make-shift door onto the dirt path again. The walk to her school was short, but in the heat it felt like we were walking for hours. We found a spot in the shade and waited in silence. It was too hot for conversation. All one could do was sit and stare. At one point, Simon shook his head, went to look for some water, found none, and returned to rest his head in his hands. I looked at the groups of silent Sudanese men in the compound. They were tall and skinny, their skin a deep black. Some had parallel lines of scarification on their foreheads, a traditional marking for men in southern Sudan who have come of age, though it seems the practice was going out of style. Not many of the men I met in the camp had it. Most of them have grown up under the influence of Western aid groups and United Nations agencies, separated from the traditions of their families and homelands. Neither of the major Sudanese rebel leaders had this scarification either, and I wondered if this influenced the practice.

As we sat, I tried to imagine what the men were thinking. They sat on benches or pressed against buildings, trying to keep themselves in the shade as the sun rose higher overhead. They looked at me and some came over to shake hands and exchange greetings.

"Hello. How are you?" they asked in English.

"Fine, thank you," I said.

"Very fine," they responded, smiling. And we shook hands again. Then someone else came and repeated the exact same exchange in the same intonations. A brief receiving line formed and the "conversation" replayed itself a few more times. When everyone had said his "very fines" we plunged back into silence. When it is too hot for action, too hot to entertain lively conversation, when no one has work and everyone is simply waiting for an undetermined event—an arrival, a departure—the days can stretch out forever, like the landscape, flat and unchanging. The morning was filled with stillness and penetrating, boneboiling light. Another young man entered the compound and we shook hands. How are you? Fine, thank you. Very Fine. Yes. We shook again.

The young man exchanged some words in Dinka with Simon, who nodded, and then the young man went to another group of men, shook hands and sat with them. We were quiet for a several minutes. Noon approached, and our shadows melted into the light. I noticed no discernible change, but suddenly something prompted Simon to speak.

"Charity cannot come today," he said. "She sent this boy to tell us she has work to do for the family." He was silent again. We did not rise to go. We waited, preparing ourselves to step from the shade into the sun. As time passed, the temperature rose. I wondered how long we would sit here inactive, neither expecting anyone to come to us nor going anywhere to meet anyone. How could Charity work in this heat? What did she have to do? A fear entered my mind that I could not possibly step into the light again, could not possibly face the heat or make it back to my air-conditioned room before I turned to vapor. I need not have worried. No one moved. No one spoke.

While we sat, I thought of the kids playing with the wooden car. It seemed so natural that boys would be tinkering with scraps, building all the possibilities of their imagination. They fled everything they knew and had to build their cars on strange soil covered in a foreign dust. Despite the violence in the camp (there was gunfire four nights in a row while I was there; one young man was killed due to a feud with a local tribe over livestock) and the chronic deprivations and stresses of refugee life, they still built toys, still dreamed up games, and still played. I considered that at the time, and still do, a triumph of life even more remarkable than those few trees that survive in this desert.

There is no easy life in a refugee camp. The total dependence on outside assistance for materials to build a home, food to eat, supplies to do anything at all, affects every member of the displaced society. There is psychological stress beyond what they survived in getting to the relative safety of the camp. Freedom of movement is limited due to legal constraints and safety issues; there are few activities and little work if you can't get a job with the United Nations or one of the non-governmental organizations (NGOs), so men become listless and bored, sometimes depressed. I had been told by many women about their husbands who were productive and gentle men before arriving in the refugee camp but after time with nothing to do, these women would complain, their husbands became abusive toward them, towards their children. In many cultures, beating wives and children in order to discipline them is the norm, though there are socially accepted rules about what is appropriate and what is not. In southern Sudanese culture, you cannot beat a child with a stick any wider than a thumb, and then only on the thighs and buttocks. And if you beat your wife without good reason—whatever that might be—her family can protest and initiate a divorce, though this is rare.

But with the deprivations of war, these social rules break down. Women, working hard to feed the children, maintain the home, get the water, clean the clothes, and tend the live-stock, all with scant resources, can quickly become exhausted, physically incapacitated, and depressed. With the collapse of the adult's world, the burden to keep society going often falls to adolescents, mostly to the young women and girls. And children who do not have the support structure of their families, orphans and other unaccompanied minors, are the most vulnerable to material shortages, exploitation of labor, forced marriage, sexual exploitation, isolation. These children have to learn how to navigate the risks around them, have to make a variety of choices for their own protection, as adults and NGOs too often fail to meet their needs, to address their concerns.

In Kakuma, the "Lost Girls" represent the extreme of this isolation.

Rebecca Maluok Mayom, a Lost Girl, told me her story, which was repeated with some variations throughout my time in Kakuma by dozens of other girls. Rebecca was sixteen years old and came from a town in the southeast of Sudan near the White Nile. She was tall, her jaw set firmly. She looked very serious and very sad. Her eyes shone all the time, the glassy look of someone who strains herself trying not to see what's happening to her, because the danger is constant and to look it in the eye at every second would drive a person mad. She is looking through her life, to some place else, some future bliss that is forever out of reach.

She was in Nairobi, where we met, hiding from a man who attacked her in the refugee camp, who wanted to take her with him, who had come after her. She chose the pseudonym, Rebecca Maluok Mayom, herself, "because it is a Christian name," she said. Her story, the story of her journey, is a common one among the Sudanese in the camp, as well as among the Lost Boys who were resettled. I heard countless variations on it, though Rebecca's was the first, on my first day in Kenya. I will let her tell it:

"I arrived in Kenya in 1994, when I was a little girl. The Khartoum government attacked my village, killing indiscriminately. They killed my parents, and I fled into the bush with my cousin. We did not know where to go, we only fled and spent several months in the bush with other children. There were caretakers there as well—adults, but there was no food or clothes and many people died, eating berries that made them sick, eating nothing at all. We went to the border of Ethiopia"—she interrupted herself, wanting me to understand clearly—"this is a country neighboring the Sudan. We were received by the United Nations, who gave us food and clothes. But this man, Mengistu, he let the refugees stay in Ethiopia and when he was overthrown, we were chased back to the Sudan."

Mengistu Haile Mariam was the Marxist dictator ruling Ethiopia from 1977 to 1991, when a coup ousted him from power. He fled to Zimbabwe, where Robert Mugabe allowed him to stay, as a favor between despots. His downfall was the second act for the harrowing journey of the famous Lost Boys, which ended with their resettlement in the United States. The story of these boys has been told countless times: After being chased from Sudan by the armies from the north of the country, the children from the south marched through perilous terrain, under regular bombardment to Ethiopia, where they thought they were safe.

Ethiopia gave them sanctuary in refugee camps from which Mengistu could use the Sudanese People's Liberation Army

(SPLA), the main rebel movement in southern Sudan, to control the local Oromo population, which was not exactly friendly to his regime. When Mengistu fell, the Gambella camp was attacked by the Oromo Liberation Front, and the SPLA leaders in the Itang Camp, twenty-five kilometers away, decided to take the children back to Sudan rather than risk losing control of the aid packages coming to their charges. The *caretakers* of whom Rebecca spoke, were often SPLA officers benefiting from the food assistance provided for the children and preparing the boys to become soldiers. When they left Ethiopia, the boys' numbers were estimated between 17,000 and 25,000. Fewer than 11,000 arrived in the Kenyan refugee camp to tell the story of their perils.

Even in Kakuma, the boys were monitored by the SPLA and were in danger from agents of the Khartoum government, and the international aid community decided it better to remove them from the camp, where they were likely to be turned into soldiers and returned to the war. Between the years 2000–2003, 3,276 boys were resettled from Kakuma camp to cities across the United States. In that same time, only 89 girls had the opportunity to resettle.

Rebecca's cousin was one of those boys, and she longed to see him again. In Sudanese culture, it is neither safe nor socially acceptable for girls to live on their own, so while the boys settled among themselves in Kakuma, the girls were integrated into families, often before registering with UNHCR. Girls had traveled with the boys in smaller numbers, according the several sources in the camp, yet they were far less visible. There was almost no record of girls arriving with this wave of Lost Boys because they had already been taken into families, some of whom treated them as their own, many of whom treated them, according to a representative from the U.S. Embassy in Nairobi, "as chattel."

"I arrived in Kakuma camp after being chased from Sudan once more. Here I met my cousin again," Rebecca said. She told me how he looked out for her because he was older, and while he was there she did not have as many problems as some of the other girls. "I was very disappointed when the boys left," Rebecca said. "The boys told us we would come afterwards, but this is not our privilege. Perhaps it is a gender gap. Girls have no powers. We would like to join our brothers. I would like to join my cousin in America." Rebecca sighed and picked at the fibers on her dress.

I finally met with Charity when she finished with school on yet another hot afternoon in Kakuma. It was a good day for her because she had been able to go to school. Her teacher told me she was a good student, intelligent, though has trouble because she, like so many girls, misses a lot of school.

"I want to go all the time," she said. "But others in the community don't like it when their needs are not met. I have no parents. I must do everything to avoid quarrels, so I miss school. Everything in the house depends on girls. Boys don't have to work at all, but if a girl resists the work, she will be beaten."

Charity was savvy. She was the only young woman I met who described herself as a Lost Girl. She knew, I imagine, what that title would evoke in me: languishing in a refugee camp without her family to protect her, forgotten by the international community who took a great interest in her male relatives and friends, the Lost Boys of Sudan. She wielded the label "lost girls" like a brand name, shorthand for her suffering and the suffering of other girls. She hoped the words might have the same magic effect they did for the boys. The High Commissioner for Refugees prefers the more inclusive term vulnerable women to describe these and other girls in the camp in need of intervention. But Charity is quite aware that the term *vulner-*

able women will not get the attention she and her compatriots need. Though living in a camp in the desert, she understands the imperatives of the media age.

"I tell the others," Charity said. "I tell them that, no matter how hard it is to tell, they must tell their story. They must keep telling it and telling it and telling it. It is only through people knowing our story that they will understand what we have been through and will help us."

Like so many children of war, she believed that telling her story would open doors for her if only the right people would hear it, would believe it. For the Albanian children in Kosovo, this was the Hague Tribunal for War Criminals and the international community who could grant independence to the province; in Lugufu camp it was the charities that provided resources and training, and in Kakuma, where everyone wanted resettlement, it was the media and the American INS, the gatekeepers to a new life over the ocean. The telling was never intended to be therapeutic; it was barter.

Charity told me the story of her journey, which was similar to (though not identical to) Rebecca's. Fleeing to Ethiopia and then back to Sudan and then, after an ordeal of near-biblical proportions, to Kakuma.

I also heard this story from Patience, Hope, and Susan. Perhaps this tale had been adopted by those left behind in the hopes it would influence their outcome. This had to be true in some cases. I tried to verify the tales as best I could, but I was inclined to believe most of the girls. The stories sounded much like other stories I had been told by other refugee children around the world. The fear, the violence, the *sounds* of violence. They rang true. While some of the details may have changed over time in the camp, during the flood of interviews the Lost Boys gave which the girls must have heard, I believe the narrative is a kind of collective memory of what each child went through, her individual story folded into the general story of Sudan's Lost Children which, by the time they spoke to me, had become legend. As the UNHCR Protection Officer told me, "Everybody knows the story, knows what to say. Everybody [who got resettled] *is* a legend."

This legend forms a large part of the way the young Sudanese girls see themselves. It is their history, like the Battle of Kosovo for the Serb children in Kosovo, it is the story that gives them a sense of identity and of purpose. The myth matters; the telling matters, though unlike the medieval battle in the Balkans, there is the hope in the telling of the Lost Girls' story that it can change their future.

"I have suffered the same as the boys," said Patience, seventeen years old. The Dinka tribe from southern Sudan are a terribly beautiful people. They are generally very tall, with broad shoulders and deep black skin. Supermodel Alek Wek comes from the Dinka tribe, as does NBA star Manute Bol. Patience fits in with this group; she's tall, has powerful arms and shoulders but delicate features. Her hair is pulled back into cornrows, and she wears a flower printed dress.

"I don't remember when we went to Ethiopia because I was very young. I was with my father. My daddy just grabbed my hand and we ran." Patience was in Panyido Refugee Camp, which came under bombardment from the Oromo Liberation Front in Ethiopia. The refugees were forced to flee again, having lost support at the fall of Mengistu. The OLF thought of Sudanese as loyal to the former dictator who had given them protection, and therefore they expelled them. The children and families found themselves back in Sudan, where the attacks by the government in Khartoum continued. They decided to head toward safety in Kenya.

"This is what I have seen. There was a lot of starvation; there was no food. Many people died in the river Gilo. They drowned; they were dragged under by animals. My father was there and paid so we could use a boat to cross. Then we were attacked again at Pochalla, where the Red Cross gave us some food. My uncle died; my older brother was wounded. My mother fled with my older brother, and I remained with my father and the younger brothers. We fled to Bor." In Bor, the birthplace of SPLA leader John Garang, the massive group of refugees came under attack again. Amnesty International estimates that, in what became knows as "The Bor Massacre," 2,000 people lost their lives. Thousands more fled the killing. Patience and her family arrived in Kakuma in 1992. Her father died in the camp. She does not know what happened to her mother or older brother (Figure 15).

"Now, things for me are very bad. You see, many of the girls were taken in by foster parents, but they do not care for them. The interest is always wealth," she said. She is referring to the practice of the dowry a family receives when a daughter marries, which, since the traditional age for a daughter to get married is fifteen, looms over the heads of young women on their own in the camp.

Wealthy men offer between 20 and 100 cattle to a family in exchange for a girl of marrying age. Amid the deprivations of life in the camp this offer is hard to resist, especially when the girl is a foster child. The welfare of the bride becomes a much lower priority. If the girl resists, she is beaten.

"The family may not tell the girl what is happening," Patience explained. "They make an arrangement with the man and then send you to fetch some water. While you are there, the man will come and take you by force, whether you cry or not; that's where your life ends."

* * *

Young women who have no parents fear being forced into marriage. In an ideal Sudanese marriage, the woman's father and brothers would act as a line of defense against abuse. The bride could go to her father if she suffered physical abuse from her husband, who is supposed to take over the role of protector. Without parents, women like Patience and Charity had no options at all.

"If you want to go to school, your husband can say 'No, fetch water now' or 'Wash the clothes,' and you must. If you complain you will be beaten. You must do what you are told if you are a woman. You must keep quiet," Patience said. Yet when asked about the dangers of rape and abduction for girls who have no parents in the camp, an officer from the United Nations High Commissioner for Refugees (UNHCR), the organization responsible for *protection* of the refugee population, responded that the threat to girls was lessened by marriage.

In preschool in Kakuma camp, the ratio of boys to girls enrolled is roughly one to one, according to figures provided by Lutheran World Federation. By secondary school, when girls are of marrying age, the ratio is seven boys to every girl. Girls at school often suffer harassment from the boys and from the male teachers, said the same official from UNHCR.

I met many outspoken young girls, caught between the traditional practices of their culture, in which women work to support the family in which they live, and the desires for education and independence they see women embracing around the world.

Many girls care for their younger siblings if the parents are not present or able to do so. While they would like to go to school, they feel obligated to protect their families. Patience, who is sixteen and therefore marriageable, fears a forced marriage more than anything, not only because it would cut short her own plans (to become a doctor) and subject her to permanent servitude, it would leave her brothers with no one to care for them. Even while we spoke, one of the little boys came up to her crying, and she comforted him, calmed him down, and sent him off to play again. "Without me, they would have no future. No chance," she said in perfect English.

Later that week, when I asked Charity about her future, what she wanted, she told me: "What you are aiming at is that you are not killed. Everything else. . . ." She waved her hand as if she were clearing a table and said no more.

Charity has had to make a lot of compromises to protect herself. Besides missing school to placate her family, she married at seventeen years old, in the hope of defending herself against forced marriage to someone else, illustrating, much to my chagrin, the UNHCR protection officer's point that girls might be safer in marriage, giving up whatever individual dreams they had for themselves.

"I did not want to," she said. "I married too soon. There were allegations in my family, they tried to force me to marry someone I did not want to, one of the boys going to the United States. I married my true love instead, though he cannot pay the dowry and was chased away. I live with my family until he can pay. I am not feeling good." She looked around the compound. We watched some women go by in bright sarongs, singing, carrying water on their heads. "I want to be free," she told me as we watched the water-carriers walk by.

On a walk through the camp, Patience pointed out the house of a girl who had been forced to marry.

"We cannot visit her," she told me. "We may cause trouble and she would be beaten."

The phrase echoed in my ears. I heard it over and over again as the days progressed in this camp. *And she will be beaten.* Like a mantra. *And she will be beaten and she will be beaten and she will be beaten.* Anger washed over me, began to tear at me. After one week in the camp, one week hearing these stories I felt a helpless rage blurring my vision. Imagine the rage of these girls, these vivacious, intelligent young women, *who will be beaten.*

In spite of their drive, in spite of their intelligence, their ambitions, all that they have to give, their lives in this place are a continuum of submission: submission to the war, to the desert, to the policies of governments and aid agencies, to their families, to the men who choose them, who take them as wives, submission to their culture, to traditions that many wish to cast off, submission, inevitably one day, once more, when they have worked their bodies to the bone, to the desert again. Patience deserved better. She deserved options in her life. So did Charity. So too did Rebecca. And how many others? How many I hadn't met? I would never meet? The rage was dizzying and pointless. Against whom, against what was I raging? This was the world. This was the world these girls came from. This was the world to which they would return, the only world they knew.

"It is very bad," Patience said, either reading my mind or reading the angry blank of my face. "Many girls do not survive this, you know?" She sighed and crossed her arms, squinting at the sun and then back at me. "It is too bad." Her idiosyncratic English said it all. It was *too bad.*

"Yes," I said, unable to find a hopeful word.

She found it for me. Unknowingly—I assume—quoting Gloria Gaynor, she shook her head. "*I will survive.*"

I asked Patience if she has other friends in similar situations to her friend we could not visit, any that I might be able to speak with about their situations.

"Yes, I have many friends like this." She went on to list the names of about five girls. I asked her if there was a way to arrange visits with any of them.

"These are not the girls who are here," she told me. I thought I misunderstood her. Her English was not perfect, so I asked her to explain. "Abduction is a problem," she said.

Abduction is a big problem facing Sudanese girls, the protection officer with UNHCR confirmed. Kenyan and Sudanese men, it is said, will arrange dowries with the girl's family and, as Patience told me, the girl will know nothing about it.

"She can be walking down the road and a man comes with his brothers and takes her, and there is nothing to do. She can kick and cry and they just laugh." Sometimes the men will take the women out of the camp, back to Sudan or into Kenya.

A seventeen-year-old girl, call her Hope, arrived with her aunt in Kakuma in 1994. Her extended family cared for her because her parents had died, she said. She went to school and did the usual chores for the family: cooking, cleaning, fetching wood and water. But when the time came for her to go to secondary school, her aunt was not pleased. She wanted Hope to continue helping with the substantial amount of housework. School was for the boys. What business did a young woman have at school? For what purpose reading and math? How would this help the family? There were fees for school. The family was poor. Why did she want to punish her aunt by leaving her to do the chores?

Their relationship deteriorated.

Two years earlier, her aunt made an arrangement with a Sudanese man for twenty head of cattle in exchange for Hope. She resisted, and her aunt hired a man to abduct her and take her to the border. She was a poor woman, and it must have seemed a sound investment to hire this man. The girl was her property, valuable property and the fee was a small price to pay to have it returned. The cattle were well worth it. Getting this expensive *schoolgirl* out of her hair was worth it too. Hope escaped at the border with the help of a local Kenyan.

"A good man took pity on me," Hope said, "and brought me back here. Now I live with another relative. I miss much school to do work for him, so that he does not force me to this [older Sudanese man]." She is, in short, at her relative's mercy in the refugee camp.

"The problems we had in the Sudan, the raping, the killing, these things are still happening to girls in Kakuma, in the refugee camp," explained Rebecca, hiding in Nairobi. For the young girls of Sudan, the war was only the beginning of their worries. They must still make hard survival choices, they are still in danger, perhaps more danger from members of their own community than from outside enemies.

I saw this over and over again: young people, especially girls, whose greatest worries came from their own community rather than an external enemy. The memories of war and violence did not plague them as much as the ongoing hardships of their lives in the camp, the prisons that their futures often became to them. The terror of the present rather than the traumas of the past.

"For myself, for the future," said Claudia, a twenty-year-old mother whose child is the result of rape by a group of soldiers in the Sudan, soldiers she had to placate in order to reach the refugee camp. Otherwise, she would have died in the desert. One can see that she was a beautiful young woman, which can be quite dangerous, but the hard years had taken their toll on her. She looked much older than twenty. Her baby nestled in her arms as we spoke, sleeping.

"I don't know what will happen. That can only be determined if I went to a good place, a safe place with my child and little brothers. How can I go back to Sudan? No one will care for me because I have a child and no husband. I make beaded jewelry to distract myself from what has happened to me, but I would like to go to school and I would like to be safe."

She told me she suffered from nightmares, but she had no one she could talk to, no one who could comfort her. She often thinks it would be better to die than to be in her situation, she said. She thought about hanging herself, she told me, but then no one would care for her baby.

She had been thrown out by her foster family for having a child and had taken refuge with another family, but, she said, there are many boys who want to take her by force and she has no one who can help her. Sometimes, she is forced to sleep outside with her baby with nothing but a small blanket to cover them.

According to an official from the U.S. Embassy in Nairobi, "there is a substantial but as yet unquantified population of women living without shelter in the refugee camp."

I spent one day at a center for young mothers and met many girls like Claudia. Most of their children were the results of rape, and several had children with men who later abandoned them. I met three other girls who had no regular shelter and were always at risk of attack by local tribesmen and by men who wanted to take advantage of them (either from their own community or one of the other nationalities in camp).

There are common features, regardless of geography it seemed, to life in a refugee camp. When I entered Lugufu camp in Tanzania, I saw a group of boys playing with a small wooden car they had made out of twigs and scraps of wood, though their car had what looked like part of a bicycle in it. Another group of boys played soccer with the ubiquitous trash and rag bundle that seemed to be the regulation ball for youth all over the country. A true World Cup Soccer Match, I imagine, would be played in the dust with an improvised ball. That's how the world plays, at least the world in which play is the most precious thing, hard won and lost too easily.

There was red dust everywhere, covering the sides of the white UN jeep, in the folds of my clothes, in the cracks of everyone's skin, in all our hair. Refugee camps are usually established on the worst piece of land a country can find, the places no one wants, at least no one of any importance to the government. Dust seems to be a universal feature. Refugee camps are the world's waiting room, its repository for the unwanted, the disregarded, and the dispossessed. This is where people go when power fails them or when it bares its teeth.

This was true in Kakuma and it seemed true in Lugufu. Rugged, inhospitable land with inhospitable neighbors. Of course it is easy to judge the host countries, seeing the land on which they put the desperate people who come to them for safety. However, some of the poorest countries in the world host two-thirds of the world's refugees. That they have anything to give is remarkable enough; that they give it is miraculous. The needy always outnumber the generous. It is true the world over, in Kakuma near the Sudan-Kenya border and Lugufu, near the Congo-Tanzania border, on the Thai-Burma border, and in the Balkans.

Another thing that was the same in any camp I visited were the problems faced by young women and girls, the threats, the forced submission, the shriveling of choices.

Jeanine was fifteen years old when we met. She came from Burundi, where her family worked a small plot of land. The war between the government and rebels forced her to flee. She wore a white dress with flowers on it. The white had faded to a reddish yellow with the dust. Her hair was braided. She looked like any number of girls I met, dressed up for her meeting with the *mzungu*, perhaps wearing her only dress. Jeanine is classified as a "street child" in the camp because she left her foster family and lived on her own without permanent residence.

She asked me very early in our conversation if I was married. As I had already had one young woman ask if I wanted a girlfriend and a marriage proposal from someone who wanted to come to the United States, I was hesitant to tell Jeanine I was single. I told her I was engaged, which was a lie.

"Oh," she said. "You have no ring."

Well, I thought, lie to one of these kids who have few advantages but their savvy, their ability to read people to get what they need, and of course you'd be caught out. I worried right off that I had negated any trust she and I might have built.

I told her she was very observant and that I was caught in my lie. I explained to her why I had lied, my (self-centered) worry that I would be put in uncomfortable positions by the truth, as I had been already.

She laughed at me and said she did not want to go to America. She wanted to go home, back to Burundi. The longing in her voice was clear when she spoke about her homeland, an ache for the land that made me think of the Joad family in *The Grapes of Wrath*. She wanted her own bit of soil, rich black soil to run through her fingers. Her soil; her family's soil.

"We had land and many types of food. It was good land to farm. We would do some farming and some trading, growing many things," she said. "We were happy. During the war, there was much fighting and I hid in the bush. When the fighting around my home stopped, I left the bush to find my parents. I found that they were dead. I saw their bodies, and I went to Tanzania with strangers. The strangers would not care for me. I met a woman who was kind and she brought me here to safety."

"Were you afraid when you arrived?"

"When I arrived, there were rumors of a bad spirit in the camp, and outside the camp there were blood suckers and animals that kill people. I was afraid then. I'm not afraid anymore. I have gotten used to living here."

"Do you feel safe here?"

"Living in the camp is too difficult. It is hard to build a home. It is hard to get enough food, and we always eat the same yellow peas. It is hard to get enough supplies, women's supplies."

Safety and supplies are one and the same for girls. If girls do not have access to sanitary napkins, then once a month they must stay out of school until their period is over. If their clothes are torn, it is impossible to go to school, as they could be seen as goading men on. Falling outside of accepted norms can be an invitation to violence against them. Non-food items are essential to the well-being, education, and safety of school age girls and are often in short supply.

"Some people go into the Tanzanian villages [to get supplies] but it is dangerous to go without permission. Sometimes you are beaten and sometimes you are not. They may rob you and leave you with nothing. You cannot go alone, and I have no one to go with me."

"Does your foster family help you?"

"They mistreat me. I have left their home."

"Your foster parents mistreated you?"

"No, there is a brother and sister. They mock me for being an orphan. They insult me so I do not stay there anymore."

"Do you know other children who have lost their parents, in school maybe?"

"No, I do not go to school. I stopped going because the other children tormented me about my past. About being an orphan." She sat with her dress tucked between her knees and looked over her shoulder out the window. "I want to learn to read and write. I would go back to school if they would not torment me. One day, when there is peace, I want to go back to Burundi and take my inheritance for my family. For now, I am getting used to living here."

Jeanine felt very alone with her problems. In Burundian culture, women cannot inherit property, and somewhere in her mind, she must have known that. The land would go to her uncles or brothers, if she had any. I didn't want to remind her that her plans were nearly impossible, because her regard for the future seemed to be a source of strength for her and perhaps she would grow up to challenge societal norms, to get her own piece of earth. She had already chosen to leave her foster family and live on the streets rather than endure their taunts. I got the sense from one of the social workers in the camp that she was a troubled girl, who made a lot of trouble for the agencies that tried to help her (specifically for this one social worker), though towards me she was polite. I think her troublemaking was not too different from many adolescents who act out to get attention, but in this case the attention she sought was to help her survive, to help her get a better break and put an end to her abuse. I had arrived in a UNHCR land cruiser, though I had only hitched a ride. I assume that seeing my mode of transport influenced Jeanine's behavior towards me. She figured I had influence and put on her best behavior. She was not at all oblivious to power relationships in the camp, knowing that a white UN officer would have some pull over the social workers to whom she had been rude.

Jeanine did not realize that there were well over a hundred unaccompanied minors living in the same camp who had experiences similar to hers. Though she felt it, she was not alone. Many other orphaned girls have faced similar problems, sometimes without anyone finding out about them.

Mathilde, who was thirteen, lived with other unaccompanied minors in another camp for Burundians. Firewood, it seemed, was the first thing on her mind. She drew me a picture of girls collecting wood. She mentioned wood in her answers to nearly every question about life in the camp. She was determined to talk about collecting wood, though she was not doing it out of a love for the task.

"I do not like collecting firewood," she said while looking at the floor. "In Burundi, I stayed home to guard the house while my parents worked," she said, puffed a little with pride. "I did not have to carry food or water or firewood."

Her father died, she told me, of disease in Burundi. Then her mother took her to the Congo, where she died of disease as well. It could have been any number of diseases that killed her parents, but given the stigma attached to AIDS, it is possible that her mother moved after her husband's death to avoid harassment within her community. This would explain why her daughter now lives outside the care of a foster family in the refugee camp. Often, children whose households have been devastated by AIDS, as Keto explained to me earlier, are shunned by the community out of fear and can find no one willing to care for them.

"On my own now, I am responsible for getting wood with the other girls in our house. Our neighbors," she said, meaning the Tanzanians, "come and attack us while we gather wood. They come and beat us and kill us." She would not elaborate any further. When asked if she had been attacked she simply said yes and fixed her gaze on her big toe. The male translator advised me not to push any further, and I could tell what he meant when he translated "beat us and kill us." He nodded to me that I had understood correctly.

Though there has been a decline in sexual violence against women gathering firewood in Mathilde's camp, young girls who are walking to remote areas are extremely vulnerable to attack, especially since they are considered "safe" targets for sexual violence because the risk of STDs and AIDS is lower. In the case of Mathilde, if she was attacked, she did not have any adult to inform in her life. I learned from a rape counselor later in the day that Mathilde had never spoken to anyone about being attacked, at least any adult.

By living with other children in a similar situation, Mathilde did not suffer from the social isolation that Jeanine felt. She went to school and wanted to be a teacher when she grew up, like many children I met. Teachers were some of the very few positive adult figures in their lives, which might explain why so many of the orphaned children wanted to be teachers. Additionally, school was a child-friendly zone in the camps, a place where they were free to be children, usually the only one, and attending was a privilege that not everyone had. In school they could play and learn and goof off. I think some of the children saw teaching as a way to stay in school all the time. Children saw that teachers got to go to school every day, unlike most of the children. They liked the stability that implied, stability that their own lives lacked. On a more practical note, teaching was one of the few jobs one could get in a refugee camp, one of the few adult employments any of them saw.

With the other children, Mathilde laughed and played with ease. She would like to improve camp security, "so there are no problems for girls," she said.

Nicole, who was ten years old when I met her, had been a refugee in Tanzania since she was six. She came from the Congo and lived in Lugufu camp with her grandmother. She had been brought to me by the same aid worker who introduced me to Keto, Michael, and Melanie, though Nicole did not know them before that day. She was extremely quiet, shy with the other children and very shy with the strange foreigner who was looking at her drawing.

"I don't know what to say to you," she said. She liked to draw. Her favorite subject was Art. She sang a song to the translator and myself instead of talking. Her favorite song, she said, taught to her by her mother. It is one of the songs women sing when they are getting water from the well, to pass the time together. After the song, she spoke a little more freely and told us about her daily routine.

Nicole went to school in the mornings, but only if her grandmother did not need her for work until later in the day, and only if her clothes were in good enough condition for her to feel decent. If her clothes were soiled or damaged, she stayed home, she said. It is not appropriate for a girl to go out in torn clothing, and as she approaches puberty, not safe either, said my translator.

At school, she recited her lessons and obeyed the teacher. She is polite without being exceptionally astute, I was told. I have met other girls a little older than Nicole who say more, who reflect more deeply, but Nicole possessed a gentleness that I found endearing.

Her face flickered with a nervous smile every time I asked a question. I got the distinct impression that Nicole was not often asked what she liked or thought. Even though she had trouble thinking of what to draw and had to be prompted to depict her favorite game—Monkey-in-the-Middle—I detected in her a creative impulse that far exceeded her resources. Her desire to communicate in song spoke to this, as did the depth of her Monkey-in-the-Middle drawing. In the picture there are three girls, no more than stickish figures in the kinds of little skirts most children draw to show what is female. One of the girls faces out of the perspective of the drawing with a downturned mouth. She is the girl in the foreground of the picture and looks quite a bit like little Nicole in crayon on paper. She had drawn a self-portrait, she confirmed (Figure 16).

What she could not express in words, she showed me with the little picture of herself in her perfect red dress. Nicole's real dress is slightly worn out and beginning to fray at the edges.

"My favorite place to be is the well," she said. I can picture her, the little girl in the picture, tossing a ball back and forth and splashing around in the water.

"I have lots of friends to play with at the well," she said. "I like the well more than any other place. In school, the boys take the balls away." Even doing drawings before our interview, the boys took half of Nicole's paper away.

"At the well," she explained, "I can play with my friends and have no worries. I play jump rope and monkey-in-the-middle, and this is my best time." She smiled proudly, unlike the little girl in the drawing.

This is not to say that girls are the only ones who have a hard time in refugee camps or that they are the only ones who have to manage their own survival or that of their friends and families. Boys face many problems as well. The lack of opportunity can make them feel inadequate, unable to make the leap to manhood that employment and self-sufficiency signify. Without much to do, they are easy targets for military recruiters, criminals, pimps, or drug pushers. Seen as adult-like, with adult capabilities (look at Johnny and Luther Htoo, look at Paul, the precocious child-soldier), boys are often sent from the camps to the cities to work on their own, support themselves. They become part of that mass of boys on the streets of cities of all the world, hopefully finding fellowship with each other, staying clear of crime and unscrupulous police officers, and surviving.

To go to the city where there are opportunities or stay in

the camp where there is international aid? To go to school and prepare for the future or to work now and have something to eat, to feed the younger siblings? These are no easy choices for young people (or adults either!). The most resilient kids I have met seem to be the ones who engage with these choices, who think critically about them, who feel a sense of responsibility towards others but also towards what they often call "the right things" or "good things" or sometimes frame in religious terms with "God."

For children who do not make it to a refugee camp where at least some measure of protection and stability is provided, these choices are all the more important.

Furaha was fifteen years old when we met in Bukavu, the war ravaged Congolese city on the lake. We met in a children's shelter in one of the neighborhoods high above the city. Groups of children ran between the buildings carrying shovels, digging drainage ditches to keep the water from tearing apart the buildings, which were all in danger of toppling due to the heavy rains and the unstable ground.

Furaha and her four young brothers fled the fighting in their village outside Bukavu in April 2000. She is a tall girl with short hair and a severe expression. She speaks very quickly. When I ask a question she always looks down at the floor to listen, pauses, looks up at me again and rattles out her answer as if it had been waiting in her all the time.

"Furaha," I ask, "how did you come to the city? What happened?"

"My father and mother were killed by the soldiers. We couldn't stay because of the violence. After my parents died, we had no home, no place to stay. We came to the city on our own,

Figure 1. Miroslaw's depiction of the Battle of Kosovo and the Death of Lazar. The 1389 battle still haunts the province of Kosovo.

Christof, a boy of mixed Croat-Serb parentage, suffered torment for his ethnicity in the years after the war in Bosnia.

All drawings and photographs courtesy of the author

Figures 2, 3. Images of school, soccer, and violence dominate the drawings of most former child soldiers.

A group of former child soldiers in the eastern Democratic Republic of Congo, some of them, including Paul and Xavier, had been there for months, unable to find relatives or a foster family willing to take them. Reintergration into civilian life is often the hardest part of recovery.

Figure 4. Keto's depiction of his escape from the war in the Congo.

AIDS is destroying the fabric of society in much of sub-Saharan Africa. One boy's drawing of "AIDS Man."

Figure 5. Melanie drew images of different things she saw when she fled the Congo. The weapons remain ingrained in her mind.

Figure 6. Justin found comfort in the rhetoric of children's rights. Here he states that children have the right to go to school and to work.

A group of Sudanese children living in Kakuma Refugee Camp in Kenya, horsing around with the author in the summer of 2003.

Figure 7. Nicholas drew the attacks on his village, which continue to haunt him. In the picture, he hides behind a tree, a tiny witness to the carnage.

Figure 8. May longs to attend school, believing that school, and only school, is the key that will open the door to a brighter future.

An Albanian boy in the town of Zahaq. Zahaq, like many places in Kosovo, suffered from the campaigns of ethnic cleansing in the late nineties. Serb paramilitary units murdered the fathers of several children I met at this boy's school.

Figure 9. Ostar's father, in better times, before they fled the civil war in Burma.

Thinzanoo and Ostar pictured with the rest of their family in hiding in Bangkok. Shortly after our visit, they were forced to relocate, due to trouble with their neighbors. Very few Burmese migrants in Thailand were granted refugee status at the time of my visit. Without refugee status, this family was viewed with disdain as illegal immigrants.

Figures 10, 11. The dream world depicted by exiled children of Burma. They are all homesick for land in their pictures.

Figure 12. Flags and guns were linked in many of the drawings by exiled children from the Karen ethnic group in Burma. They are tied up in a sixty-year-old nationalist struggle.

Figure 13.

Aung Su, a Karen boy from Burma, lives in exile in a small town near the Thai-Burma border. His favorite hero, Spider-Man, cannnot protect him from the harrassment he sometimes receives at the hands of Thai teenagers.

Figure 14. Aung Su's full-color drawing of Spider-Man.

Figure 15. The harrowing journey of the Lost Girls of Sudan has become a modern-day epic in the oral history of those in Kakuma Refugee Camp.

This group of girls living in Kakuma Refugee Camp in Kenya came from southern Sudan, where the long war has robbed them of their parents. Unlike the famous "Lost Boys" these girls have little hope of resettlement in the United States, and many face forced marriage and a life of servitude, in spite of their ambitions to become doctors, nurses, and teachers.

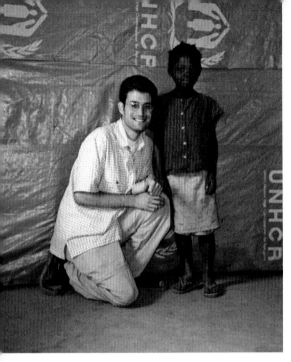

Nicole, a shy girl living in Kakuma Refugee Camp, loves to play Monkey-in-the-Middle. Even though she is very young, she has witnessed a great deal of violence and has lost much of her family.

Figure 16. Nicole's drawing of her favorite game, Monkey-in-the-Middle.

Figure 17, 18. Robert, a street child, drew his dream of a house and of safety, though he also drew a kung fu fight, in case anyone wanted to mess with his dream house.

Figure 19. Musa's drawing of a soldier shooting a man in the back of the head.

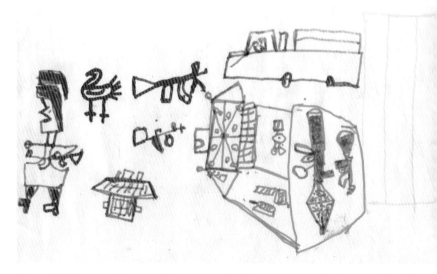

Figure 20. The generic images of weapons are far from generic to the former child soldier who drew them. They each represent a specific weapon or aspect of his training.

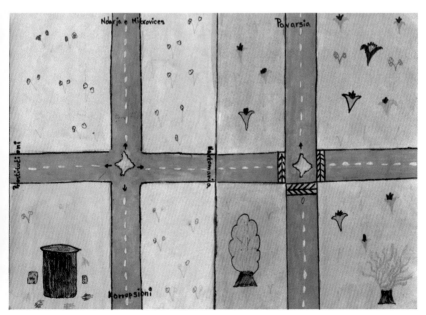

Figures 21, 22. The children of Kosovo look at the country as it is, and put all their hope in independence to give them a brighter future.

One of several outdoor schools in Lugufu Refugee Camp when I visited in 2001. The war still raged in the eastern Congo and new children arrived in the camp every week. Resources were always in short supply for the children who wanted to learn.

Figure 23. Bujana's house, ringed in barbed wire, reflected the feeling common among Serb children in Kosovo that they are trapped, penned in, and in danger from the Albanian majority.

walking the whole way. It was a long journey and we were very tired. I led my little brothers here. When we got to the city we wandered the streets and met some people who told us about a center where they helped children. That is where we went, and they found people for us to live with. Strangers."

"Do they take care of you and your brothers now?"

"One of my brothers went to another shelter, but I am still responsible for all my brothers. I feed them and raise them because there is no one else. I have big problems. I worry that we will have no place to stay; that the strangers will not take us into their home anymore. We have no place of our own and I have to care for my brothers somehow. I learned how to write, but would like to learn more skills. I want to make dresses and earn a living, to care for my family."

"How do you earn a living now?" She looked down at her feet, the same way she did for the other two questions, but the answer came up short and slow after a very long wait.

"I get money here and there. I get help from the center, some." She closed her mouth and did not speak anymore.

"What do you want for yourself for the future?"

"For me? I only want to make dresses and earn a living."

An aid worker told me later that many young girls in the city who are trying to make a living and stay off the streets will become prostitutes, selling themselves to men for, in some cases, as little as a small meal. When I told the aid worker about Furaha she told me that the sweet and well-spoken fifteen-year-old I met was probably selling her body to survive, though she could not be certain. For children in the war-strangled city, this is one of the few options open. A local NGO representative listed the choices facing children in the eastern Congo: "to join the military, to become a street child, or to die."

I spoke with a large group of street children at another children's center one rainy afternoon. They were brought to me from the market by a man who runs an organization that tries to look after their needs for schooling and clothing. He told me it would be too dangerous to go to the market myself to find the kids. It would cause too much commotion and draw a lot of unwanted attention to me.

He arrived at the center in a swirl of loud and rowdy kids, all excited to be picked out of the market and given lunch, crayons, a ball to kick around, and the ear of a foreigner. This certainly did not happen to them everyday. Many had never met a *mzungu* before.

It was hard to discuss very much with any one child, as all the others would gather around to use the crayons or listen to the conversation. They liked to weigh in about their favorite movies, laughing and joking about the best parts: a cool gun battle in *Rambo*, a funny dismemberment in *Starship Troopers*. All the films were violent, and all ages attended the screenings. As we talked about Bruce Lee and some fantastic kung fu moves, I looked around and realized that twenty boys surrounded me. Among all of the kids who had come that day, not one was a girl. I asked the caretaker about this.

"The girls are usually taken in before they come to us," he told me.

"Taken in by whom?" I asked.

"Either the army or other men." He told me what the aid worker had suggested. Young girls on the streets of this city have little opportunity outside of prostitution and the army and the pimps are quick to grab adolescent girls, taking advantage of their adult-like physiques and their desperation. There are few jobs for the adults, almost nothing for younger people. Many young people choose to sell themselves or go into the army because this increases their chances for survival. The pimps and the army recruiters seem to be the only options available, so children come to them. They are more than willing to put that youthful energy to use.

Robert did not know his age. He thought that he was fourteen, like the boy next to him, but he said he cannot be sure. He was separated from his parents and lived on the streets of Bukavu with no idea of their whereabouts, whether they were alive or dead. The scabs on his skin, which he could not stop scratching, could be cured by a bar of soap, "but we have no soap," he said.

"During the Kabila war"—meaning the war in the Congo that began when Laurent Kabila overthrew Mobutu—"we ran away from Goma with my parents. But I got lost and separated from them. There was a boat leaving, and I wanted to get on it, so I asked a woman to please say that she is my mother if they ask, because I have no money to pay for the boat. She helped me, pretended to be my mother, and I came to Bukavu on that boat." Like most of the street kids in Bukavu, his story of homelessness began with war. Uprooted and far from home, he had no family network to look after him, only the rag-tag wisdom of thousands of other street kids who have fled to Bukavu as well.

"I met some kids when I got here and asked them where to sleep. They told me about a place near the market and let me follow them there. I spent time with them begging, but only during the day. I do not beg at night because people don't like it, only in the day. I never steal. I try to stay away from bad people." I noticed a package of cigarettes in the pocket of Robert's worn out shirt.

"I am too young to smoke," he told me. "I sell the cigarettes in the market one by one to make money."

Manu, a lean and hungry-looking young teenager had more trouble in the market than Robert. "At first, I did not know where to go. I slept in an alley and was obliged to steal because of hunger. I lived for a month like that. At night, the police came and demanded money from me, because they knew I was begging. But I didn't have any money, and they beat me." A local organization that helps street kids told me that this occurs regularly. Often, they say, street children are arrested for no reason and held in prison, where they suffer abuses and malnourishment. Human rights groups pay handsome bribes to get access to these kids and monitor them, so holding street children can be a very profitable business for unscrupulous policemen. And if the human rights groups don't know about the kids or cannot pay, there are always the pimps and soldiers.

Robert had his own problems with the police, he added. He described having a job for a while that gave him the means to eat and sometimes to see one of the videos that play in the movie houses. *Rambo* and Bruce Lee, he said, are his favorites. Everyone around agreed. When there is no money, he and the other children tell me, they wait by the butcher in the market and eat what falls to the ground, sometimes raw meat off the bone.

"My job was calling out for passengers to fill a bus," Robert said. "I don't have this job anymore, because the man with the bus—the man who gave me the job—was arrested by the police and they sold his car. He crossed the border without permission, they say, so now I need a job again." He picked up some crayons and drew a picture of himself begging, the sun shining overhead.

"Once the war is finished, the street children will be finished. We can all go home." He showed me his last drawing, a colorful house where he used to sleep (Figures 17, 18). The other kids looked at it with smiles. They liked the kung fu fighting Bruce Lee in the picture, because "Nobody can mess with him," they said. "He can fight all the bad guys."

* * *

The lives of the boys on the street in Bukavu and the lives of boys in the refugee camp in Tanzania did not seem all that different to me.

Justin, the Rwandan boy who liked to talk about children's rights, told me about his life in Lugufu camp:

"In the camp, it is suffering. When we sleep, we are four people under the same blanket. We are given a jerry can and one basin to bathe in. These are our possessions."

I was told by an aid worker that Justin probably had his own blanket taken by the foster parent or foster siblings he lives with, as sometimes happens to children who live with a family that is not their own. "I've gone through a lot in the camp. There are good people in the camp. My neighbors are good. They give me food. But it is hard. When I sit and cry, no one comes to help me. When I fell sick, I got no attention. I had a fever, and I had to walk to the hospital. I carried a boy who was wounded. I used to be healthy, but now . . . the environment here is bad. Sometimes people don't use the toilets."

Keto described similar hardships:

"Because of war, we don't have people to care for us. Our education is harmed. In the camp we don't get everything we need, especially when compared with what elders receive. We don't get enough clothes, enough soap. This can cause disease."

Just like the boys on the street in the eastern Congo, the boys in the camp worried about cleanliness and also about "bad people," who would judge them or try to hurt them.

With Keto, I met a group of boys his age who complained that other children made fun of them because they had no parents, that local children taunted them because they were refugees.

"Maybe the other children hate me inside, though they don't say it," said Abwe, who the other children told me was a good goalkeeper in soccer. They helped him with his drawing because he said he didn't know what to draw. Keto helped him much of the time while I interviewed other children. Toward the end of our conversation, he said that "people don't like us; they boo at us, say this is not our place. We don't have anybody to take care of us, but people are always calling us in to eat with them in different places. I want stability. I have two neighbors who are not good people."

I asked him how he knew they were not good people.

"I have a feeling they are not good people. They eat everything and don't leave any food. They insult and abuse us."

Abwe kept returning to the abuse he suffered as a refugee. Like many of the boys I knew to be living on their own, he talked about "we" and "us" when telling about his problems.

Samuel, two years younger than Keto, added, "You can see how shabby I am. I used to dress sharp. We are suffering. We left everything behind. In your country there is no war, but if there was war, you will also flee and lose everything that you have."

Samuel was embarrassed because he had lice.

"Because of war," Keto said, "we don't have people who will care for us. If a person is not a refugee, he should not look down on refugees." The others agreed.

Adolescence is a complex series of navigations, even in the easiest of circumstances. People start treating you differently because you look like an adult but you still feel like yourself, and then your hormones go crazy and you aren't sure what you feel half the time. You have to redefine yourself constantly, in your family, in society, even in your own mind. For adolescents affected by war and displacement, a range of complex choices and definitions is added to this mixture. The Lost Boys of Sudan had to learn to cook and clean for themselves when they lived on their own, which was unheard of in their culture. Keto, in Lugufu camp, had to choose with whom he should go after his mother died, where he should stay, where his chances were best. He had to earn money to pay for secondary school, which he could just as easily not have done, sitting around in boredom like many young men or working and not returning to school like many others. For him, school was the option that would give him the most chance for survival.

I heard the phrase "Education is my mother and father" over and over throughout East Africa and have read it in several re-ports by other researchers. Keto's regard for the future (he wanted to be a driver or a mechanic), his decision to value education as a goal worth striving for, kept him going in the present, just as Jeanine's hope to go back to her parents' farm kept her go-ing. Their hopes might be unrealistic, even futile in the face of continuing wars and insecurity, but having those hopes for the future helps them to survive in the present.

It was the children I met who had hopes and plans who did not fall into the depression that life in the camp can bring. Claudia, the young Sudanese mother, had no hopes or plans. She did not play an active role in her survival, beyond passing the time with beadwork. She was waiting for someone else to help her, to rescue her and take her away from the suffering. She became depressed because her chances of being rescued were slim to none. She had fallen into a despair common to people stuck in refugee camps but truly terrible when it afflicts someone who is still young. Not all young people are capable of great acts of will, of harnessing their survival instincts into creative energy.

Lepaix, a gangly young teenager, told me he was five years old. My translator had trouble understanding him, told me that the boy was having trouble communicating. He looked closer to fifteen years old than to five.

"Do you remember when you got here?" I asked.

"Last year."

"Do you remember how you got here?"

"There was a war." He paused. He looked around. His head moved slowly, floating on his neck. He did not seem agitated, engaged in our conversation, or interested in being anywhere else. He seemed to be elsewhere already.

"And you came here because of the war?"

"I heard war, guns. I saw others running. I ran too."

"How did you cross the lake?"

"A crowded boat."

"Did you know anyone on the boat?"

"No."

"Do you know where you parents are?" This question was always hard to ask. It seems insensitive and foolish when speaking to war-affected children living in a camp on their own, but much of the work done by non-government organizations, in association with UNICEF and the Red Cross, is tracing unaccompanied minors to family members in the hope of reuniting them. Sometimes, I have been told by Red Cross employees working on tracing, a child says both his parents are dead though it turns out one is alive somewhere and the child can be reunited with family. The search is not always successful, but any information can help, so, after some time with the child, I ask the question. I worried when I asked Lepaix that it was too soon, that I asked only because I was frustrated that our conversation wasn't going anywhere, that I wasn't getting enough material. I was projecting my frustration onto him, deciding he was a bad subject to be interviewing. I wanted to be done, to get the facts and to get out. It is only now, looking over my notes, that I see, besides the likely developmental problems he had with language and body-weight—he was very scrawny—his disengagement was itself telling, and I should not have thought of it as my failure as an interviewer or, even worse, his failure as an interviewee. He had much to teach me had I been more patient, and I regret not spending more time with him.

"My parents died of disease . . . I was told, when I was young," he said.

"Where did you live after that?" I asked in an overly official tone.

"With the others . . ." His voice, which had been lingering near silence since we started talking, faded completely at this point. He did not look at me or the translator, but he did not seem to actively avoid looking at us either. His eyes still drifted.

"Who are the others?"

"The others who worked in the mine with me."

"You worked in the mines?"

"Yes."

War was destroying traditional structures in the Congo. My translator explained to me that normally a child whose parents had died would be taken in by others and cared for, but this child had been working in the mines since he was very young. It seems no relatives could take him in or no relatives wanted to, perhaps because his parents had died of AIDS, perhaps because he had developmental problems, perhaps because the burden of one more mouth to feed was simply too much.

When talking about the mine, his attention came to us and he stopped drifting. He focused on me, and, while still having trouble communicating clearly, he was more animated and engaged with the game of questions and answers.

"I was paid 503 Zaire francs." He smiled. This was barely a

few cents when the country was still Zaire, at least five years before we met, when Lepaix must have been around nine or ten years old. Lepaix told me the mines were very dangerous, though it was good to have work to live and friendship with the other children his age working there.

He talked about the diamond mines more than anything else. He spoke of the hard work, the damp conditions. It worked like this: A boy, the smaller the better, is lowered on a rope to the bottom of a 100-foot hole. The boys need to be small so they can maneuver in the pit. They dig out clumps of dirt and throw the clumps into sacks, which other boys haul up and take to the river, to sort through, looking for diamonds. The boys spend all day at the bottom of the pit. If they have to relieve themselves down there, they do. The men who exploit them, the men who make all the money off these operations, have no interest in the welfare of these boys—young boys, hungry and alone are commonplace in the eastern Congo, their lives are cheap.

"At times, when digging, there would be accidents," Lepaix said. "There would be a collapse, the dirt falling in, crushing. I have seen others killed this way. The mines were dangerous."

He smiled a bit, for no clear reason. At the start of the interview, his cheeks were flushed, and his first instinct was to answer every question put to him with a simple, "I don't know." By the end of the conversation, he had grown somewhat more spontaneous with his statements, though still quiet.

I asked him where he would go if he could go anywhere in the world. He said he didn't know. After suggesting a few generic places that he could visualize—the city, the ocean, another country, home, he still said simply, "It doesn't matter."

I asked him what he likes about living in the camp, what he liked about where he came from.

He told me that life in the camp, where he lives with a "kind old man" and goes to school, works, and plays soccer is the same as life in the Congo and in the mines. "There is no difference, though I had friends in Congo."

He thought about home. Living in exile, he missed the only life he knows. When the war ends, he said he would go back and work in the mines again.

"Where will you live?"

"With my friends. The others. We will be together and take care of each other."

He had been on his own for a long time. He had learned to go along, to work, to do what he was told. He was the object of a lot of outside forces: the mines, the bosses, the fighting, the refugee agencies. But it was in the mines that he got some satisfaction, it seemed. Working in the mines he had earned a salary, had formed his own bonds of friendship, had survived dangers, perhaps numbed to them. In the camp, he made no plans, formulated no specific ideas, seemed barely aware of his surroundings. His emotional distance protected him, perhaps, from terrible thoughts of the things he has seen, but it also kept him from taking a more active role in his future. He could not have his hopes disappointed, because he did not seem to have any.

Looking back now, I find his situation quite troubling. He survived as if on autopilot, easily submitting himself to exploitation by the mining concerns who knew that they would not have difficulties controlling him. The isolation of this adolescent boy is dangerous. His work in the mines, the sheer number of child laborers in the Congo, and the large number of girls engaged in the sex trade shed some light on the assertion made in a Women's Commission report that "adolescents' strengths and potential as contributors to their societies go largely unrecognized and unsupported . . . while those who seek to do them harm . . . recognize and utilize their capabilities very well."

This report, *Untapped Potential*, examined the myriad ways adolescents can and do contribute to the healing and recovery of societies torn apart by armed conflict. It also looked at how few programs supported their potential and how sometimes the design of aid focuses solely on very young children and adults. These sorts of programs alienate adolescents and push them further toward the recruiters, the thugs, and the pimps.

Without systems in place to protect and engage children like Lepaix, vulnerable children and orphans, there is no shortage of unscrupulous adults who are ready to exploit them. Not all children are capable of resisting or escaping to find other choices.

I cannot say I would have the strength or the will of Keto or Patience or any of the kids I met who get up every day and face their confines and dangers with courage and a degree of optimism, who cope with their grief and help others to cope. I can see myself wanting only to be rescued, waiting and knowing that my waiting is hopeless or allowing myself to be used by others because it made things easier. Growing up Jewish, there were, inevitably, times I couldn't help but imagine how I would have fared in a concentration camp. I wondered how I would have held up to the mental and physical strains of that assaulted life.

Until I met these kids in Africa, scrawny and certainly not in the best of health, I thought I could never have survived such situations alive, let alone with any of my current values intact. But listening to them, I feel I might have learned something. Under strain and stress, one doesn't necessarily have to vanish, to die mentally or physically. As George Orwell wrote, " . . . many of the qualities we admire in human beings can only function in opposition to some kind of disaster, pain or difficulty." Many of the children I met illustrated this perfectly. I found much in them to admire, much that had been forged by disaster, pain, and a great deal of difficulty.

It is a commitment to themselves, to their families and communities (sometimes despite the abuse inflicted on them by the same), and to the future that brought out the best in the young people I met. Young people are quite capable of making commitments to their communities, their moral sense, or their beliefs and acting on those commitments when they are able. If adults can recognize and encourage this, adolescents, even in the most difficult circumstances, can be valuable forces for good in their communities. Ignored, these same capacities in children can be harnessed by other adults as powerful destructive forces.

FIVE

"The Things I've Done"

Children as Soldiers

The former child soldier sitting across the table from me had little hands. They wrapped around the crayon with a delicacy that defied the thoughts I could not shake: those little fingers had pulled the trigger on a Kalashnikov; those little fingers had shot to kill. He wore a jean jacket that was too big for him and his hair was cut close to his head because of a lice problem. It made me think of a U.S. marine.

There was no more essential image of war in Africa in the late twentieth and early twenty-first centuries than that of the little boy clutching a machine gun taller than himself. Currently, 40 percent of armed groups around the world use child soldiers. Twenty percent use child soldiers under ten years old. In Sierra Leone child soldiers, hopped up on a mixture of heroin and gunpowder, cut off the limbs of civilians, sometimes their own families, sometimes other children. In Colombia, children executed their peers for infractions such as falling asleep on duty. In Gaza, terrorists packed children with explosives and compelled them to turn their bodies into "holy shrapnel." Many families took great pride in their child martyrs. In the Democratic Republic of Congo, commanders forced new child recruits to take part in the ritualized cannibalization of prisoners to complete their indoctrination. This is what I knew about child soldiers: violence, blood, and terror. And here was one such little boy sitting with me, taking great care of his drawing. He took his time selecting colors: reds, pinks, and oranges.

The boy's name was Musa, and he came from the Ituri district of the eastern Democratic Republic of Congo. Somehow, he found his way to Bukavu through the jungle. High up the hill in one of the poorest districts of the poor city, we sat at a table in the back of a small house belonging to a charitable organization that aided demobilized child soldiers. The house was crumbling. Rather, the hill on which the house sat was crumbling, falling out from underneath the building. Mudslides were common in this neighborhood, and they took a little more of the buildings with them each time. In order to enter the structure, we had to leap up to the first step. I imagined in a few weeks time, a few more weeks of rain, this leap would not even be possible. Yet the children at the center and the adults working there laughed at their building that was sliding off the hill. They had survived worse, I imagine. They laughed, though they wondered where the children would go when the building met its final flood and in a neat somersault, tumbled over itself down the hill, crashing through the battered city to the lake.

Musa told me he was fifteen, as Paul told me, as they all told me. I sighed, sure he could not be fifteen, and jotted down in quotation marks "15?" I had spoken with two other child soldiers that day, both of whom said they were fifteen. I had spoken with a few the day before. All of them fifteen years old, most of them looking younger. I was beginning to tire. Their stories left me drained, getting to know them was becoming emotionally difficult for me, as I knew I would have to leave them and would probably never speak to them again. The stories themselves were hard to listen to.

There were soldiers just down the road from the center, maybe one hundred yards away, and they were always watching, making notes on who went in and who came out. If they wanted, they could snatch a child back into the army as soon as he got home, or they could make sure he was harassed or threatened until he rejoined. One of the most pernicious aspects of the use of child soldiers is that, in the drawn-out conflicts in which they tend to be used, the cycle of violence continues and, unless they are carefully monitored in the long term, the children are regularly re-recruited into the army, caught and punished for escaping, or even recruited into the opposing side.

In defense perhaps from getting to know Musa too well, I began to plug in assumptions about the boy based on what I had heard from the others and what I read in my research, rather than listen to another tale of terror and cruelty transforming a child into a warrior. He had not yet started speaking.

Musa was probably walking home from the market or from school or playing soccer with other boys when a truck drove up beside him. Roads and schools are two of the most common places for children to be abducted. Rebel groups in Colombia, Sri Lanka, Myanmar, Congo, Uganda, Liberia, Sierra Leone, Angola, Sudan, Nepal, Indonesia, and terrorist organizations operating out of Pakistan target schools as recruitment centers as a matter of policy. Anywhere that there are a lot of children away from their families, however, is ideal. Orphans are the best, because no one will miss them and they have no hope for themselves anyway. I sighed and assumed Musa was an orphan.

The soldiers in the truck would have forced him to go with them, grabbing him and tossing him in the truck or tricking him by offering him a ride. Then, he would have been taken to a camp somewhere remote or to an airfield and loaded onto an airplane to take him to a base in the jungle. That's where they'd train him: first how to march, how to obey orders, how to clean for the adults, how to handle a wooden gun, how to shoot a real one. With the training would come abuse, verbal and physical. Merciless. The cruelty inflicted on him would desensitize him to inflicting cruelty on others. Then they sent him to the front, goaded him on to commit atrocities, to kill and to maim, or they assigned him to the most dangerous jobs, running as a decoy through enemy lines, picking up ammunition off the dead, walking through minefields, spying. Either way, they demanded that he kill, so that he would be baptized in blood. Or they demanded that he rape someone, baptized in sexual power. Maybe he contracted AIDS. As Physicians for Human Rights observed in Sierra Leone, 50 percent of rape victims tested positive for HIV, and male child soldiers were considered the primary transmitters. Perhaps Musa became too sick to keep fighting, so they let him go rather than care for him, or maybe he got tired of the army and, when the commander wasn't looking, escaped. I steeled myself to hear this story and began with the first question which I was sure would lead to this terrifying and predictable tale.

"Can you tell me, Musa, how you came to be in the army?"

"My friends beat me up and for revenge I joined the army," he answered without hesitation. "I volunteered because I was angry."

"You volunteered?"

"I went to the recruiters and they took me in. I joined the army to have power over my friends, but I got no revenge." He sighed with disappointment, shaking his head at what a foolish child he had been a year ago. "They took us to a training camp at Luama and made us get up early every morning and do exercises. Then they moved us to Mushaki Base and the commander took me out because I am too young."

This story did not fit my profile. Not at all. Musa volunteered. He was not kidnapped, not forced to be there at the point of a gun. He *wanted* to be there for some kind of preemptive revenge. He never fought in the bush, though he wanted to. A commander took him out of the army and sent him to an organization that works with former child soldiers. The commanders were supposed to be the bad guys, the children the unwitting victims of manipulation.

"I disliked being in the army and I want to go home," Musa said. He drew a picture of himself playing soccer and himself sitting in school at the rehabilitation center. He drew a man running from a soldier delivering a stream of bullets into the back of the man's head (Figure 19).

Musa was growing up in the eastern Congo at a time when societal rules had disintegrated. The *interhamwe* and various armies and militias roamed the land, raping, killing, and looting. The norm for government officials, police officers, and soldiers was (and still is) corruption. It is impossible to know how many violent acts Musa had witnessed in his life or what effect they had on him, but according to his story, in the army he never saw combat. Yet his picture contains some graphic details of violence, resembling the pictures of children who have witnessed such terrors. The drawing could be a product of his imagination or a depiction of things he had heard as easily as it could be an eyewitness account. The point is that the image of a soldier wielding absolute power over a weak civilian was in Musa's mind. With violence so prevalent in his young life, it makes sense that a feud with his friends, the kind of quarrel children have all the time, would lead him to join the army as an impulsive response to his anger. What he knew about the world he lived in—that soldiers have the power to get what they want and to act with impunity—led him to believe that being in the army would make sure his peers could not harass him again. His decision to join seemed rational from his point of view.

Not all children are forced into military service. While children's recruitment cannot exactly be called voluntary, various surveys have found that over half the children fighting in most forces joined under no direct threat of violence. Other factors, such as societal shame, perceived security, poverty, or a lack of job prospects may have made military service the only option in these youth's minds, but, like Musa, they were not abducted. They went because they wanted to go. They believed that the military would get them what they needed, whether it was power, revenge, money, food, protection, validation, or simply something to do. Recruiters all over the world, from the United States to Myanmar, know this and take advantage of it. In Sri Lanka, playing on youth's desire for adventure, recruiters will show up at schools with a motorcycle. The glitzy army recruitment commercials on U.S. television during prime time play on the same feeling. Sometimes, armies will take in unaccompanied children, orphans, and migrant workers, telling the youths it is for their protection, they are not forcing them into the army. But the youths rarely stay out of hostilities for long.

This made sense when I thought about some of the unaccompanied minors I met in refugee camps. They did not want to be helpless recipients of aid, they claimed. They wanted to become responsible and productive adults. Keto, for example, wanted very much to prove his own self-sufficiency and competence. I can imagine that, had he not made it to refuge in Tanzania but instead had been brought into the army, he would have "volunteered" to fight so as not to burden his unit, to prove that he too could contribute. And others, whose wills were not as strong as Keto's—Lepaix for example—could easily be persuaded to fight alongside the grown-ups.

Some of the exiled Burmese children I had met might have joined armed forces for ideological reasons. I know at least one of them has done so since we met. Children are certainly not without ideology. As Robert Coles discusses in *The Political Life of Children*, politics do get worked into the life of the young. Children integrate ideologies into their views of the world, as evidenced by racist dogmas spouted by young people in South Africa during apartheid or communist doctrines espoused by children involved with the Nicaraguan Sandinistas in the 1970s. In Iraq, children are fighting the U.S.-led forces that they see as occupiers of their homeland. Coalition forces have engaged with soldiers as young as thirteen. As one adolescent boy who fought against U.S. forces in Najaf in 2004 told the London *Daily Telegraph*, "We will kill the unbelievers because faith is the most powerful weapon."

In situations of protracted armed conflict, the opportunities for children to participate in hostilities are many, and it is only through a stringent effort *not* allowing them to fight, that youth can be kept out of military service. During the Taliban's reign in Afghanistan, Mullah Omar passed strict edicts against using child soldiers, even though they had been extremely helpful when fighting against the Russian occupation years earlier. Systems of punishment were established for commanders who used children who could not yet grow a beard. For a short while this worked, but in order to keep their forces strong against the Northern Alliance and the feudal militias who used children, the Taliban gave in to the temptation for cheap and easy labor, enlisting eager youths in scores, trucking in refugee children from sympathetic madrassahs in Pakistan.

In the absence of a strong will to keep children and adolescents out of the army, history proves, they *will* join. During the American Revolutionary War, children found ways to fight. Hezekiah Packard enlisted at age thirteen after the rebel victory at Bunker Hill. Some boys, prevented from joining because of their age, argued until the recruiters, who had quotas to meet, let them join. During partisan resistance to the Nazi occupation of Poland, Jewish children joined in droves.

The choice was simple: die in a ghetto or concentration camp or go down fighting. The resistance needed these fighters and used them eagerly.

In the eastern Democratic Republic of Congo at the time I visited, no political will existed to stop youth from fighting. All parties to the conflict were actively engaged in recruiting children as soldiers, porters, and sex-slaves, even though the 1999 Lusaka Peace Accords called for the disarmament of child soldiers.

When presented with evidence of minors below fifteen years of age in their ranks at the front lines, the head of the recognized government in Kinshasa's Department of Foreign Affairs did admit, as Human Rights Watch reported, that they had "inherited" child soldiers recruited by the late President Laurent-Désiré Kabila, for the campaign he waged together with Rwandan forces against the Mobutu government in 1996 and 1997. It took several years for them to begin to demobilize some soldiers, and even though it was largely a cosmetic gesture.

In December 2001, Kinshasa demobilized 3000 of its child soldiers. There was a large event as the children ended their service in the armed forces—UNICEF provided T-shirts. In response, the major rebel party in the east, RCD-Goma, committed to demobilizing 2,600 child soldiers. UNICEF did not send T-shirts. RCD-Goma demobilized only 140 of the 2,600 children. The rest were moved from the Mushaki military base forty kilometers outside the city of Goma to remote locations in the jungles of the Katanga province, far from the eyes of the international community. Could this have been a response to the slight by UNICEF for not delivering T-shirts or was the RCD simply unwilling to let their young recruits go? Some aid workers have indicated that most of these demobilizations are a sham. They only let the weakest or most troublesome young soldiers go. The undesirables. This rings true when one considers that very few girls ever participate in these demobilization events. The commanders want to keep their spoils.

Driving into the area controlled by RCD-Goma in January 2002, I saw an armed border guard who could not have been older than fifteen. A few days later, I saw the boy again and asked my guide, who worked with an NGO trying to get kids out of the army, if this boy was a holdover from before the Lusaka Accords who had yet to be demobilized since the RCD signed and made a commitment to demobilize their child soldiers or if he was a new recruit.

"That is difficult to say," my guide told me. "I know this boy. He will be fifteen in May. He was a soldier before Lusaka, but he was demobilized in our center, given shelter and food until we could locate his family or place him somewhere. We found his family and he rejoined them. The army took him back a few months ago and he is a soldier again."

Authorities have acknowledged and sought to justify the continued use of child soldiers. In an interview broadcast on January 24, 2001, on the RCD-Goma's Radio Goma, journalists asked Commander Obert Rwibasira of the RCD-Goma's G5 military division why the movement continued to enroll very young recruits. He replied that RCD-Goma needed a "young and dynamic army."

The Uganda People's Defense Force (UPDF), which has deployed not only in northern Uganda to fight the guerilla Lord's Resistance Army (LRA) but also into the eastern Congo, admits that it has underage fighters in its ranks, even fighters who have escaped from the LRA. "If somebody at seventeen years comes from the LRA and takes the choice that he wants to be in the army, would you send him away so that he returns to the rebel ranks or you help him become productive?" Shaban Bantariza, the Ugandan army spokesman said. "You let him return to the bush, which he has known for most of his life—or the lesser evil of taking him while slightly underage and give him a chance to change his life?" he added.

At the same time as the Ugandan army seeks to justify its use of child soldiers, despite being a signatory to the UN Convention on the Rights of the Child (which forbids using child soldiers), Bantariza denies that it is their policy to do so. "Last year we got thirty recruits who had been duly recommended by the community councils, but after scrutinizing them [we found] they were underage and their applications were turned down," he said. It seems the Ugandan army only recruits children who were previously abducted by the enemy, suggesting that they are damaged goods, that their protection as minors is less important than the protection of those who had not been abducted by the LRA already. Bantariza noted that there were few rehabilitation programs in northern Uganda that dealt with child soldiers and those NGOs operating in the region were already overstretched, preferring to take in younger children. Whatever the justification, the government of Uganda, far from protecting children from the threat of abduction by the LRA, continues to use youth on its front lines. Uganda's commitments to international law do not seem to outweigh the benefits of using child soldiers. This is the modus operandi for most groups who use child soldiers. Public commitments matter little when there is a war on.

While RCD-Goma claimed to be demobilizing all of its child soldiers, recruitment continued. In December 2001, a month before I met Paul in the demobilization center, RCD soldiers burst through the doors late at night. The raid happened quickly, and the armed soldiers ignored the objections and pleas of the civilian staff. When it was over, one hundred children had been taken back to the Mushaki military base. The staff showed me the bullet hole in the door and the broken lock. They complained that there was little they could do to protect the boys in their care if the army wanted them.

In 2005, an aid worker in what used to be an area controlled by RCD-Goma said that, as the fragile peace in the region deteriorates, military commanders beef up their ranks with children. "[The Mayi Mayi] have realized we want the kids, so they won't give them to us," the aid worker said.

The fighting in the Ituri region, which displaced over 80,000 people in the first four months of 2005, involved children ordered to maim civilians believed to be of rival ethnicities. Local defense forces—sometimes calling themselves the Mayi Mayi— also made use of children to protect against these attacks. While commitments to protect children are easy to make, breaking the habit of relying on young fighters is much harder.

One boy, Augustin, told me about being taken by force into RCD-Goma. He told me he was sixteen years old. He was more likely fourteen years old. He said he was recruited when he was thirteen and spent one year as a soldier. He had only been at the transit center for a few months. It is likely that he was instructed to say he was fifteen if anybody asked when they took him into the army, again, because fifteen is the absolute minimum age allowed by international law for participation in the military. Now that a year had passed he said he was sixteen, improvising on his instructions.

"I was forced into the army," Augustin said. "The soldiers from the RCD passed by my house. They gave me supplies and told me to carry the things for them. After I carried supplies for them, they told me to go to the airport with the things. At the airport they instructed me to get on the plane with them to go to Goma. There were a lot of soldiers so I went. When we got to Goma they said, 'You are no longer a civilian, but a proud soldier.'" Then they took him to Mushaki training camp and began teaching him to fight along side other *kadogo*, little ones. Girls and boys trained alongside each other at Mushaki, learning to march, obey orders, and shoot. As one boy told Anne Edgerton, an advocate with Refugees International, the child recruits "were fed once a day, the food generally consisted only of porridge, they had to sleep outside in the rain and they were beaten."

Xavier was abducted into the army in the middle of a soccer game. He was with his friends kicking the ball around when a group of soldiers pulled up in a truck. He demonstrated when we played in the courtyard, pointing out where the soldiers stood, casting the production with the other kids—and interrupting the main narrative with an essential detail, how he could juggle the ball for five kicks without it hitting the ground. The other boys in our soccer game played along for a moment. They took the role of his friends, also taken, or of the soldiers, cool as can be, rallying the children into the truck, making fun of Xavier's soccer ball juggling bravado. Paul, whom I had already interviewed, played one of Xavier's friends. He did not want to be one of the RCD soldiers. Musa, who had volunteered, played an RCD soldier with confidence, adding a swagger to his step and lowering his voice to a baritone growl, which made the others laugh.

Many of the children knew this story all too well. Though they had fought in different armies, many of them had similar experiences, were taken the same way. All over the world, soccer games are interrupted by the realities of war and violence. The kids knew the parts to play in this scene, they played them ferociously well. There were moments with the shoving, with the role-playing soldiers when I wanted to intervene, but the moment passed quickly, the soccer game resumed, and Xavier returned to the bench where I sat, where the others could not hear, and continued his story.

The men forced the kids into the back of the truck where they sat with two soldiers who had guns. The soldiers were from the RCD. I could well believe that Xavier, the gentle soccer player, didn't want to go with them, was reluctant to become a warrior. Though picturing how he charged into danger as soon as he had the ball, I could also imagine him hopping in the back of a truck with a kind of youthful carelessness, looking for adventure. Regardless of his motivations at the moment, the soldiers had guns and did not offer the children a choice. They drove all of the children into the forest to begin their training. "At the training there were five girls and nine boys. We were all sent to the front line."

Xavier looked at me and crossed his arms on his chest. He said something to the translator with a nod of his head at me.

"He wants to know if you want to hear about the battles," the translator asked.

I was thrown off. It seemed like such a painful subject to discuss. I was ashamed to say yes, I did want to hear about the battles. How does an adult ask a young person to talk about such horrible things? What kind of person wants to hear it and moreover, wants to write it down? I felt like a voyeur, but listened intently.

On the front, they clashed with the Mayi Mayi, the militia

with which Paul had fought. In the rehabilitation center, the boys got along quite well. On the day Xavier described, they may have been shooting at each other. I was glad the other boys were absorbed in their soccer game again, showing no interest in our conversation. Maybe they already knew each other's stories, maybe they didn't want to know.

Xavier and his unit were patrolling through the jungle when machine gun fire cut through the trees. Xavier spoke in a level voice as he described the battle. He uncrossed his arms and leaned forward, looking at his hands on the table. He began to pick at a splinter in the wood.

The guns crackled against the trees. Everyone was shouting. The other children were afraid. Everything went silent for a moment and then the air came alive once more with fire. The commander yelled orders at the kids. Whenever one of them left the cover of trees or bushes, he was mowed down by the enemy guns.

Suddenly, a barrage of bullets tore through the commander, the one adult among the youngsters. It was fear of this adult that kept them in battle. Without him, Xavier felt no reason to stay in the hell of machine guns blasts and shouting children. With his commander dead, he escaped. He ran away from the bullets and into the mysterious jungle that he had never been in before. He knew only that whatever he found would be better than his chances in the army.

"I have killed many people, I think, but I don't know. I don't count. It is better to forget those things. If I could speak to my recruiters now," he said, "I would tell them to study and learn, not to become soldiers. I suffered very much."

"What happened to the girls in your unit? You said there were five?" He looked up at me surprised, as if the answer was obvious.

"All of them died," he said.

Girls are used in the army for a variety of purposes. Michel, a sixteen-year-old at the center who may have actually been sixteen, told me that he and his sister joined the army because their parents were dead and they had no jobs.

"I went to the front line many times and my sister was sent to the enemy to be a spy. Girls were sent to be prostitutes and get information from the enemy. This is how my sister was used. She is still there."

Michel had seen dozens of battles. He listed the locations of fighting in which he participated against the Mayi Mayi and the *interhamwe* in the provinces of North Kivu and Ituri, sites of some of the most intense violence in the war. He drew pictures of weapons and spoke with terrible accuracy about their technical details for such a young person with so little formal schooling.

"This is a Kalashnikov rifle and this is an R4 5.56mm gun and this is a 60mm mortar. This is an RPG." He pointed each part of his drawing out, and though to me they looked generic enough, in each of them he saw a very specific object with a very specific meaning to him (Figure 20). I was reminded of Keto and his use of terms like repatriation, transit centers, and rationing. Nothing magical about this boy's use of technical terms for weapons. They were the words that made the world in which he lived.

A Human Rights Watch report on the use of child soldiers in the eastern Congo describes the use of child soldiers at the front lines, and from this I could imagine what Michel had been through, though he did not want to talk about the battles:

[The children] were trained on how to use arms and how to shoot, and that was the end of it. Some of the kids were even sent to battle without arms. They were sent ahead of battle-ready troops of the RCD and RPA to create a diversion. They were ordered to make a lot of noise, using sticks on tree trunks and the like. When they succeeded in diverting the attention of government troops, that is to say when they drew government fire on their unarmed elements, these units, known as the Kadogo Commando, would be literally allowed to fall like flies under government fire. The experienced troops would then attack the government troops when their attention was diverted to the Kadogo Commando.

"During the fighting," says Michel, "I left the front lines with the commander." They went to sleep by the roadside on their way back to Mushaki base. Early in the morning, when the first birds were starting to sing, Michel looked around. His commander was still asleep. He stood up, left his uniform behind so he looked like a poor peasant boy, and ran away. He did not have a clean getaway.

The same men who had recruited him into the army recognized him when he arrived in his village again. They demanded that he return to the army. He feared that he would be beaten for trying to escape. The recruiters told him that they were ashamed he was such a coward, that he had made a promise that he was breaking and that that was wrong. They told him they would punish him themselves if he did not return. Michel gave in and returned to the army.

He was finally demobilized officially when Save the Children intervened and lobbied on his behalf. Save the Children works with army commanders and recruiters, educating them on the rights of children and the responsibilities of the army under international law. They lobby on the policy level with the United Nations as well as on a case-by-case basis for children like Michel, who said that when he grows up, he wants to be " a driver or a mechanic. A civilian." But this lobbying is hardly successful. What do the recruiters care about children's rights when they have quotas to meet and money to earn? As Dr. Peter Singer, author of *Children at War*, points out, most armies already know that using child soldiers will be frowned on by history, that they will be judged for doing it. Why else would belligerent parties that use child soldiers constantly deny or seek to justify doing so?

Michel's sister was still in the army, and her chances for survival were slim. Very few girls are released from armed groups that use child soldiers. They are kept as domestic workers and "wives," or are reluctant to leave of their own volition because, as victims of rape, they fear becoming pariahs in their own communities. If they have had a child while with the military group their ability to escape is hampered. Given the physical and psychological trauma of rape, the exposure to sexually transmitted diseases, and the harsh conditions of life in the bush, I imagine not many of the young women survive, or if they do, can reintegrate back into society. In some cases, a sense of duty to the families they have formed with their commanders holds them captive.

In my time at various recruiting centers, I never saw, let alone talked to, a demobilized girl, though I had seen one in uniform, still serving in the army, at the border when I entered the RCD territory and another at a checkpoint where I was stopped. Worldwide, from Colombia to Sri Lanka, there is near-total denial that girls serve in combat as more than "bush wives" and that they have unique needs that need to be addressed when the fighting stops. Of the 130,000 combatants processed by CONADER (Commission Nationale de Désarmement, Demobilisation et Réisertion), the national institution in charge of disarmament, demobilization, and reintegration in the Congo, only 2,000 females have participated in their programs.

All indications suggest that, despite the establishment of more and more international norms against using child soldiers, the practice continues and spreads. The use of child soldiers in West Africa spread from Liberia to Sierra Leone to Guinea to Ivory Coast. In the Great Lakes region it has spread through Rwanda, the DRC, Central African Republic, Angola, Kenya, Uganda, up to Sudan, down through Zimbabwe and into Congo Brazzaville. And this trend is not isolated to Africa. In the twentieth century, child soldiers have been used on every continent except Antarctica. Right now there are an estimated 300,000 child soldiers around the globe, and given the continual flare-ups in violence in the Middle East, that estimate is no doubt low.

To fight this trend, several resolutions have been passed and non-governmental organizations have been working with state armies, rebel groups, and militias to make them aware of the harmful effects of child soldier use and the norms of international law protecting children. Passing laws and sensitizing commanders to children's rights is hardly a solution. The militias know they are operating outside the legal and moral fold. Often, they make their money in the illegal trade in drugs, diamonds, or other natural resources. Often they exist solely to make money from these illegal trades. They also know that no one is capable of enforcing any law against them, because, in the regions where they use the young soldiers, they are the law. As Dr. Singer says, "You cannot shame the shameless."

In March 2004, a mentally handicapped sixteen-year-old Palestinian was arrested in the West Bank town of Nablus as he was about to blow himself up at a checkpoint. He was, no doubt, coerced into that attempted attack. In Sri Lanka in 1996, the Tamil Tigers sent hundreds of children in attacking waves to overrun the Multavi military complex. Out of a defending force of 1,240 government soldiers, the attackers (children in concert with adult units) killed 1,173 people and took the base. In Afghanistan, the first combat casualty inflicted on U.S. forces came in January 2002, from the gun of a fourteen-year-old sniper. The United States has detained at least six Afghan and Iraqi insurgents under the age of sixteen. There is no count of how many children Coalition forces have engaged in battle and killed.

Children can be a real threat in combat, one that is unavoidable on the modern battlefield. They think less about consequences; they act more rashly than adults, are less risk averse, and because they are children, can cause confusion and hesitation in the enemy. They are also cheap to train and easy to replace; thus, commanders can use them recklessly, as distractions, as forward attack units, as attacking waves to overrun enemy positions or send them in to combat zones to do the most dangerous work.

In Karbala, Iraq, during the 101st Airborne divisions assault in the spring of 2003, two embedded reporters recounted incidents that illustrate the tactical challenges and the confusion that children create on the battlefield. During the firefight for the city, according to Mathew Cox, of the *Army Times*, Pfc. Nick Boggs, of Bravo company was in position on a rooftop with an excellent view of the city. All day, the company had come under heavy attack from machine guns and rocket-propelled grenades. From this vantage point, the twenty one-year-old from Petersburg, Alaska, saw an Iraqi man sprinting for cover. The man held an RPG under his arm. From the rooftop, the American forces opened fire and the man was taken down. An instant later, two boys "no older than ten" darted from an alleyway. They went to the fallen man. The Fedayeen often sent children into battle to retrieve weapons and those weapons were used to kill American soldiers. Pfc. Boggs had the kids in his sights.

"I didn't shoot. I didn't shoot," the soldier said. When the kids reached down to retrieve the weapon, Nick Boggs had no choice. He fired his own weapon and, when the smoke cleared, both boys lay in the street, dead.

In another incident on the same day, *U.S. News* correspondent Julian Barnes reports that, during the fighting, Sgt. Jason Sypherd and Sgt. Troy Hanner watched as two boys raced out to retrieve an RPG from a fallen soldier. Sypherd shouted to his unit that it was kids in the street. He yelled for them not to pick up the weapons. Then he fired warning shots to which the boys, well trained, did not react. They reached for the weapon, and both Sypherd and Hanner fired, killing one boy and sending the other running away. A moment later, Jason Sypherd threw up. He was twenty-four years old and had just killed a child.

In both cases (which, because of the chaos, could have been the same incident reported differently), the soldiers involved hesitated in combat when faced with children on the battlefield. They had not been adequately prepared to deal with facing young children in the enemy forces. After the fighting, the soldiers involved were demoralized and depressed. "I keep trying to think of something else," Sypherd told Barnes. "But I can only think of that boy. War is a bitch."

Though they can be skilled and fierce fighters, disciplined or ruthless, child soldiers are still children who will, if they survive, become adults. For youth who have grown up only knowing war, it is believed that violence will be a way of life. Child soldiers will have missed most of their education; they will lack many life skills. If they are not reintegrated into society, given a reason for hope, given opportunities for the future, it will be all too easy for them to return to violence, whether in the service of a militia, a criminal gang, or on their own as freelance bandits. Looking at the spread of violence throughout West Africa after the Liberian civil war, one can see what happens if the former child soldiers are not given other opportunities. Young fighters from Liberia and Sierra Leone have destabilized the whole region, spilling across borders to take part in other wars. West African child soldiers have been found as far away as the Congo. The question of what to do for child soldiers when a war ends is critical, not just for the child's well-being, but for the security of the entire region.

In many contexts, former child soldiers are feared by the adults to whom the youths return. Like most adults, they believe that children who have committed terrible acts will have internalized violent behavior as normal. They have been in a terrible war (for wars that use children are by that very fact terrible) and have done unspeakable things, and therefore, it is believed, will be tainted and distorted by violence. They will have become the violence they have inflicted on others and will live in a moral vacuum. I cannot count the number of times in reading the literature on child soldiers that I have read the phrase "moral vacuum." I have written it myself countless times.

This view, however, does not take into account the fact that children who are forced to fight do not usually have a choice. Or rather, they have an impossible choice. Many of the young soldiers I met had a strong moral sense—Paul for example. He recognized that some actions were "bad"—looting and killing civilians for example. The incidents of nightmares and social problems among former child soldiers is evidence of the moral struggle going on inside them.

After Xavier finished painting the picture of his battle for me, I asked if he ever thought about it when he wasn't being interviewed by some white guy. He looked at me and laughed. "Yes, sometimes," he said. He told me that he still had nightmares about his time in the army. "All the suffering," he said. He kept picking at the splinter on the table. "I have bad dreams about the things I've done."

He did not want to tell me what he'd done, but I could imagine. He told me he was nervous that women would see him and punish him for "the bad things that I've done to women." He leaned back when he said that, no longer picking at the splinter. He crossed his arms again, but this time said nothing else. He was not eager to continue this line of conversation. It seems that Xavier had been a rapist.

It is not that Xavier lacked moral discernment. He had very much internalized the moral repertoire of his society in which murder was wrong, elders should be respected, and those who have been wronged in life exert a force on their violators even after death. He referred to both the translator and myself as sir. He, like Paul, showed great regard for the other children in the demobilization center.

He believed in a moral code that did not spontaneously appear in him after he got out of the army, just as Johnny and Luther Htoo believed in the Christian moral code even as they waged their guerilla war in Burma and took hostages in Thailand. Xavier, the Htoo boys, Paul, and countless other child soldiers simply did not have access to life choices that reflected the moral sense they had.

Xavier would have been murdered himself if he had not carried out the wishes of his commanders. Paul had no love of violence, even though he was sent to the front over and over again. He had to kill because the situation demanded it of him. It is not the internal realm of a child soldier's psychology that turns to a "moral vacuum," it is real circumstances around him that necessitate violating his fledgling moral sense. Looking at all those photographs of child soldiers that run in papers whenever violence flares up on the African continent, it isn't the children that upset me. It is what their use as soldiers suggests about the world of adults surrounding them that is so unsettling.

Child soldiers are seen as such a great threat to society in part because they undermine accepted roles for children. In war, an armed child holds power over the civilian adults. This throws off any sort of comfortable power dynamic. I learned this lesson firsthand one rainy day.

I had just left the center where I met Musa and Xavier and Paul. My translator and I drove near a market that I had been told to avoid, the one where many of the street children in Bukavu lived. It was in a bad part of town, they said. I wondered, foolishly, what made one part of town in an impoverished war zone worse than another part. I was looking out the window at the women selling nuts and the boys ambling past the car with their goats when I noticed we had stopped. I turned to ask why we had stopped, and that was when I saw the soldiers.

The car was surrounded by five soldiers. Only two of them looked like adults. The oldest, a wiry man with a mustache and mirrored sunglasses, stood next to the driver's side window and spoke in a firm voice. I have no idea what he said. The driver, Philippe, handed him his papers, which the commander did not want. He spoke angrily and Philippe responded. They stared at each other in silence, and I felt a lot of eyes on me. The soldier next to my window had a baby face and a handgun. He wore a beret on his head, and his uniform hung off his shoulders. It was far too large for him. I guessed he was sixteen. He caught me looking at his handgun, and I quickly looked away. Philippe and the officer started talking again, and this time there was anger. Philippe kept his hands on the steering wheel. My hands were in my lap. I thought about my bag, with all the names and notes. A lot of people might have trouble if these guys got hold of my notes. The commander was yelling at Philippe, who, for some reason, was yelling back. Without warning, the commander snapped at the young man next to me, who opened the car door. For a moment I wondered if he would drag me out.

He got in.

Philippe looked at him a while and then turned back to the commander and spoke very quickly. There was a rapid back and forth. The other kids around the car, all of them with Kalashnikov rifles, had their fingers on the trigger, but they were not yet pointing them at us. In frustration, the commander waved his arm and turned away. The youths raised their weapons, pointing them at us. There was a small crowd in the market watching this. Their presence helped me relax. Surely they wouldn't gun us down in the middle of the city, in the middle of a crowd, I told myself.

Philippe turned to the boy in the car. They had a brief exchange. After the longest minute of my life, the commander came back and yelled at Philippe some more. Philippe responded calmly and gestured at the market. The commander stood in silence again and then, as quickly as the incident began, it ended. The boy got out of the car, and they waved us forward, through their checkpoint and on with our day. I asked Philippe what had happened.

"Congolese people," he said. He shook his head and laughed, though it really wasn't very funny. "Everybody wants to steal the money."

They wanted to rob us. The commander used the children under his command to intimidate me. When he really wanted us afraid, he turned away and left us under the control of the children. It was only for a moment and there was a crowd around, but it was terrifying nonetheless. Had we been somewhere more remote, our escape might not have been so easy. The boy got in the car, Philippe later told me, to direct us to another place to go to continue the discussion.

Anyone who has traveled in a war-affected area knows that life and death are often decided by a simple word or hand gesture when stopped at a checkpoint, and these checkpoints are often manned by children. Philippe, a buoyant twenty-four-year-old with four children of his own, had a way with words and kept the situation from getting out of hand. He stayed calm and deferent and found the magic words that sent us safely on our way. Six months after I left the Congo, I learned that Philippe had been stopped at a similar checkpoint. He tried to talk his way out of it then too. They shot him six times.

The terror child soldiers inflict on the populations they control does not go away when they are out of the army. Most people believe that former child combatants are likely to get involved with criminal activity after they are out of the army, even though there is little evidence to support this belief. Jane Lowicki and Allison Pillsbury of the Women's Commission for Refugee Women and Children note in their report, *Against All Odds*, that in Kitgum, in northern Uganda, there have been several cases of adolescents who commit crimes and claim to be former child soldiers. They are then turned over to the rehabilitation centers who learn that they are young people from the community, not former child soldiers at all. They use the fear and guilt surrounding the Lord's Resistance Army as an excuse.

The Acholi people in northern Uganda have been plagued by the Lord's Resistance Army for years. The LRA has abducted around twenty thousand of their children and forced them to fight. When they escape and attempt to return home, they are often ostracized, resented, and sometimes abused. Community leaders express a desire to forgive them for atrocities they have committed, but they fear the bad deeds have infected the children. They must perform cleansing rituals, which free the child from the guilt that attaches to them and demonstrate that the child has sought forgiveness. Without these rituals, it is believed, the spirits of those who were wronged by the children will punish the entire community. In order for the cleansing ritual to succeed, the child must show remorse for his or her actions. Even though they did not commit these deeds by choice, they are held responsible for them and often blamed for all the crime and misfortune that befalls the community after their return. On the other hand, the cleansing ritual can allow the child to take responsibility for the atrocities she has committed and free her of the guilt.

The ritual is called *mato oput*. Many children who fought with the Lord's Resistance Army and have now returned home participate in the ritual, which is led by the elders of the community. The child crushes an egg to symbolize a new beginning; he leaps over a stick of bamboo to symbolize the leap from the past. He drinks a bitter brew made from the herbs of the *oput* tree with the people he has wronged, both parties accepting the bitterness of the past and vowing never to taste such bitterness again.

Similar rituals are practiced in Bosnia, Sierra Leone, Mozambique, and Angola. Involving the entire community in the process of forgiveness can help restore faith in social structures for everyone involved. By submitting to the will of the elders, the children show that they still respect their community; by asking forgiveness the children acknowledge the resentment the community might feel. The community is not devoid of responsibility in this problem. One of the reasons the child soldier problem remains in many areas is because the situation in the community has not changed. Violence and poverty remain. The world the children return to when they leave the military is the same world they left when they joined.

Paul, sitting in the demobilization center, was beginning to grow resentful. His community failed to protect him and then they were hesitant to take him back. This resentment can grow, and often does among former child combatants. The community must make the child feel safe again in order to rebuild trust. The adult world betrayed former child combatants, and anger at that betrayal can lead to further violence. Unless adults prove that they can be trusted not to fail the children again, the likelihood of recovery and reintegration into a peaceful society is unlikely. This is even harder when the fighting is ongoing, as in the eastern Congo and Uganda.

Sakundi was fourteen years old when we met. The vast structure where he lived was one of the best I'd seen for ex-child combatants. Sturdy gates kept the army recruiters out. There were job training programs, a school, a sewing room where girls could learn a skill and make products to sell. There was even plenty of room to play soccer, which seemed to be every boy's favorite activity. I found myself playing soccer with yet another group of former child soldiers. I wasn't much better at it than I had been a few weeks earlier, but they enjoyed laughing at me, making sure I took the ball in the face a few times.

"Header!" they would shout in English, convulsing with laughter. It was a real soccer ball and it stung.

I sat down on a bench in the shade when I got too tired to keep playing. One of the social workers grabbed Sakundi from the game and told him to go over and talk to me. He obeyed without objection and came trotting over from the dirt field.

He never complained that his game had been interrupted, though I felt guilty that my presence had disrupted his play. He assured me that he didn't mind as long as I promised to play more when we were done. He smirked a bit. He was a smartass, but a charming one. I accepted his offer, my cheeks stinging in anticipation.

Sakundi had been in the army for two years, and he looked it. His features were hard and his arms were muscular, though thin. He answered my questions as if they were orders being given. He talked without emotion about his time in the army and his life afterwards.

When he was twelve, his family sent him to the market. He saw a truck with soldiers around it talking to a group of young people.

"Anyone who wants to join the army can, because we want soldiers. Come join us," they said. So at twelve years old, Sakundi joined them. He got in their truck and left for the military base.

"I didn't tell my family because I never said good-bye."

My pen hesitated on my pad, and I looked at him for a moment.

"Nobody forced me," he said, anticipating my question.

At the military base they woke early in the morning to march and to learn how to fight. He saw older women that the soldiers used for prostitution, he said, but he had no interest in them. He was more interested in the fighting and the adventure. He wasn't interested in girls yet. Despite the adult statements he made, he was still a little boy in the army. He didn't have any friends, he said. "Only soldiers." He was learning to use a gun and to follow orders, which he liked.

"Did you fight?" I asked.

"When we heard the Mayi Mayi, our enemy, in the bush, we'd shoot at them, they would shoot at us. We would try to kill them." He acknowledged his experiences, made no effort to deny them, yet he did not want to dwell on them either. "I hated the Mayi Mayi because they lived in the forest," he said.

"One day, I was just sitting at our camp, and the commander came and took my gun and said I was too young. They sent me here to be demobilized, but I don't like it here. I want to leave."

"What do you want to do?"

"Right now, I want to find my family or study. I would like to join the army again, but because I'm too young now, I can't. One day I will again."

Sakundi's day may have come sooner than either of us thought. The day after we met, the eruption of Mount Nyiragongo ravaged the city. I had to evacuate and it was only from Kigali, Rwanda, that we could start making phone calls to see who was safe and who was not. Lava surrounded the center where Sakundi lived. We learned on the news that around half a million people had been displaced, most crossing the border into Rwanda just hours behind us.

When we got the priest who ran the center on the phone, he was with a few of the boys who stayed behind, digging a trench to protect the center from the lava flow. The other children had been sent with the general refugee exodus across the border to Rwanda and shelter was arranged for them there. I never got to ask about specific children, but in all the shifts back and forth, it would have been easy for Sakundi to slip away. If he wanted to, he could find an army willing to take him in, willing to send him back to the forest to fight his enemy.

I never found out what happened to the children I played soccer with at that center. I like to picture them digging trenches with the priest, protecting their home, and returning to school, but the jungles outside the city are still dangerous, militias still attack villages, and a Kalashnikov costs about as much as a goat. For youths who want to fight, there are plenty of opportunities. For youths who do not want to fight, opportunities are scarce.

As long as war continues, armies will need young people as fodder for the cannons. Those who choose to exploit children in this way must be punished. The companies that trade with these armies must be punished. The cost in punishment for using child soldiers or supporting those who do must become so astronomical that it is no longer worth it. More important, children must be given other choices by the adult world. As they all said, they would like to go to school instead of fighting.

School. Over and over again they said it. *I want to go to school. I want to study.* From orphans the world over one hears the phrase: *education is my mother and father.* School is the great creator of childhood, the defining space in which the pupil's relationship to the world is clear: he is a student; he is a child. In choosing school, in longing to be part of that space, that world, these former soldiers were choosing to change their relationship to society. Regular schooling for adolescents is still elusive in the eastern Congo. Young people are needed to contribute to the family economy or must fend for themselves for survival. They are in competition with the adults for what little money there is. This is a region where one in four children die before the age of five. I am reminded of the medieval concept of childhood—don't get too attached to the young, because they probably will not survive. If they do, then they work when they are able.

School, however, provides another option. In the West, children stare past the windows of their schools, longing to get out. In much of the rest of the world, children stare into the windows of the schools, longing to get in. The alternative to schooling is the harsh world of adulthood, the fierce game of survival. School not only creates opportunity for these young people; it creates their childhoods.

The former child soldiers I met had thought as adults and fought as adults, ruled over adults by force, and now they wanted to put away adult things and become children once more. Perhaps the new government in the Congo can foster peace and stability, can deliver on the promises of politicians for more schools, schools for everyone, can give these little grown-ups the chance to turn into children.

Without the chance to go to school and become children again, without hope, without opportunities, Paul, Musa, Xavier, Sakundi, and myriad other child soldiers all over the globe, despite wanting very different things for their lives and coming from very different backgrounds, are all left with the same option: to pick up a gun and to fight.

SIX

"Surviving the Peace"

Coming of Age in
Post-War Kosovo and Bosnia

How do you know a Serb from an Albanian?" I asked, kicking the ball gently to Katja, who sailed it back to me, right through my legs.

We had finished our history lesson, our story of the Battle of Kosovo, and I wanted to learn more about how these kids, five years after the war in Kosovo, three months after the riots, thought about ethnicity. Slobodan Milošević was on trial in the Hague; a referendum would soon be considered on whether Kosovo would achieve independence from Serbia. These were Serb children with whom I played; children who were frightened of being cut off from Serbia, children who were penned in to ethnic enclaves for their own protection from the dominant ethnic group, the Albanians. I wanted to know what they thought of all this ethnicity. How do you know who's who?

"They have a different language," Katja said when I returned with the ball, having chased it almost to the main road out of Lapjo Selo again.

"And they hate us," Marko added.

"You hate them too, though, right?" I asked.

"Yeah," the children answered together. "But only because they hate us," Stefan explained.

"But if your only difference is language, how do you know who to hate if they don't speak, if they stay quiet?"

"They have a different religion," Katja said without needing a moment to think about it. "They have mosques; we have churches." She was growing weary both of my poor soccer playing and my infantile lack of comprehension on this issue that to them seemed so clear, so ingrained that it need not be explored. Hadn't I heard of the Battle of Kosovo? They had just told me the story. Had I forgotten already? Did they need to tell it again?

"You can know us," little Miroslaw, the author of the drawing that started our conversation about history said, "by our damage."

"What do you mean?"

"You can know that we are Serbs by the damage we can do if any *shiptar* come here looking for trouble." The word *shiptar* was the pejorative for Albanians. It carried the same weight as *kike*, or *nigger*, or *spic* would in America. No one was fazed by it.

"If they come here looking for trouble . . . ," Marko said, smiling and acting out some kung fu moves on Miroslaw. The other boys quickly jumped in, laughing and mock karate chopping.

I turned to Niko, my translator. He was a Serb who worked with the NATO forces, disarming the local population, manning checkpoints, going on patrols. He spoke Albanian and Serbian fluently, though Albanians didn't trust him because he was a Serb and Serbs didn't trust him because he worked with the Americans. The Americans bombed the Serbs in 1998, ending the campaign to cleanse Kosovo of its Albanian population and sending the Serbs scrambling for shelter, from the bombs and the Kosovo Liberation Army, the former guerilla force that seized the country thanks to the bombing. Politics limited his friendships rather dramatically. Niko wasn't prejudiced, though. He didn't care who was Serb, Albanian, or American. He just wanted to study computers and get on with his life. Watching the boys kick and chop at each other, he shrugged and shook his head a little.

"They are very isolated here," he told me. "They don't know any better."

I watched them roughhouse, the soccer ball sitting idly in the field where my last kick had gone astray. They chased each other around, and I imagined them playing a game like Cowboys and Indians. In Israel it was Israeli and Palestinian. In Rwanda, Hutu and Tutsi, in Belfast Taig and Prod. Every society in the world has these games of otherness. As I watched the children playing, however, and thought how casually they wrote off the Albanians, how difficult coexistence was proving to be, I thought about another soccer club I had read of, Red Star Belgrade.

Red Star Belgrade was a prominent soccer club in Serbia. When uprisings against Belgrade's rule of all the provinces of Yugoslavia began in the early nineties, a nationalist thug and wanted gangster known as Arkan drew supporters for his paramilitary group, the Serb Volunteer Guard, from the fan base of Red Star Belgrade. He was a soccer fanatic; he knew the guys who hung out in the bars on game days, who gathered in crowds and went out looking for trouble with the opposing team's fans. These were his people. They loved Red Star Belgrade as much as they loved their people, the proud and oft-maligned Serb people. Their violence was born of love, of a kind. Soccer and nationalism, as any World Cup observer notices, are never far apart. That kind of love, that kind of loyalty turns easily to violence. What started as a fan club for Red Star evolved into a gang of thugs and turned into much worse.

Arkan's Tigers, as they were called, ended up acting as a death squad throughout Bosnia, Croatia, and Kosovo during the nineties, playing a role in the Srebrenica Massacre, the siege of Sarajevo, and in the campaign of ethnic cleansing throughout Kosovo. During the latter conflict, Arkan also owned a soccer club, Obilic, which spent one season as a championship team, until the Union of European Football Associations banned them from European competitions because of the team's connection to war criminals. Games and politics, politics and ethnic warfare, soccer and murder, they all merged in the Balkans.

Young Marko's popularity and charisma seemed suddenly ominous given this sordid history of play. In Kosovo, the mass graves from Serbian offensives in 1999 were still being unearthed. I had just come from the Albanian village of Lubeniq, where Serb paramilitary groups killed over eighty villagers in less than a month. The name Arkan still caused fear among the Muslim population and shame among many Serbs, though he had been assassinated in 2000. There are others, though, who see Arkan as a hero, a true patriot, and one in a long line of martyrs for Serb freedom.

Marko doing his karate chops against imaginary enemies was fighting all the enemies of history, the enemies who took his home in Pristina in 1998, the enemies who burnt houses and monasteries three months earlier, in March, the enemies who took all the jobs, all the money, the enemies who took his parents' pride, his nation's pride, his people's pride, the enemies who took Lazar's kingdom. There was no shortage of supposed enemies.

We played soccer, they roughhoused, they fought ghosts. I

feared then, I fear now, that with the wrong leadership, a new Arkan in their midst, this same group of boys could be turned on real people, the other, the enemy, the Albanians and the whole bloody conflict would start again.

I pressed them to consider their beliefs.

"But why would they come here looking for trouble?" I asked when Miroslaw drew close.

"Because," he said, in perfect imitation of Marko, "it's the history."

For the Albanian children in Kosovo, recent history had a firmer hold on their minds than the ancient battle of Kosovo. Yugoslavia had been a prosperous communist state until the early 1990s, when the republics of Slovenia and Croatia declared their independence from the Yugoslav federation. Bosnia was the next to declare its independence, which led to the deadliest conflict in Europe since World War II. Only a few years later, in 1999, the small province of Kosovo, located inside Serbia, became the site of increased violence. The province is the birthplace of the Serbian Orthodox Church, though it is 90 percent ethnic Albanian today. An organization committed to the independence of Kosovo, the Kosovo Liberation Army, had begun using aggressive terrorist tactics to agitate for freedom. The crackdown from the government in Belgrade was brutal and swift, attempting to cleanse Kosovo of its entire Muslim population. The international community responded with bomb attacks on Serb military and government sites and Kosovo, though still technically part of Serbia, became a NATO occupied territory, administered by the United Nations. Ethnic tensions between the Albanian Muslims and the small Serb population left behind remain near the boiling point.

Girls like Nora, from Zahaq, and Leo in Lubeniq, Albanian

children whose parents were killed by Serb paramilitary units, or like Eric who lost his home and his neighbors, did not need to go back hundreds of years to find the wrongs done to them. They held onto the memories of the recent civil war to sustain their sense of self, their difference from the Serbs. The province was dotted with monuments to fallen KLA soldiers, carved in black stone, and to several children who were murdered by the paramilitary units. Their names were engraved in the sides of buildings, in ubiquitous political graffiti, on gravestones.

"Serbs are different from us," Eric said, "because they speak a different language and because they show no respect for Albanians."

It sounded familiar.

Perhaps my mind played tricks on me. Did this Albanian boy look just like Marko? Probably not, but in my new turn as amateur ethnographer, trying to see the difference between Albanian and Serb, I could not. The boys were interchangeable. They were boys—snot-nosed, cocky, charming, stinky, mad as hell, mournful, stumbling graceful boys defining themselves against their enemies. Defining themselves against each other, because this is what teenage boys do.

"They speak a different language," Peter said. He was ten years old, originally from the city of Peja (called Peč by the Serbs, renamed when the Albanians took over), but he fled into the mountains during the war. Now he lives in Rugova, a stunning, but isolated region. The air is chilly and crisp. It was sweltering down in Peja, so I was glad to escape into the hills, where life is perhaps harder, but, as Eric observed, there were no Serbs around to bother them.

Peter showed a remarkable awareness of the arbitrary nature of ethnicity in Kosovo. "If I spoke Serb instead of Albanian," he said, "I'd be a Serb." He thought a moment more, not content with his answer. It didn't sound right. The look Eric gave him spoke volumes as well, perhaps pushed him to reconsider his position. "If you want to be an Albanian, you are an Albanian. It's in you."

Malesora, a twelve-year-old girl who joined us after school in one of the chilly classrooms, added her ideas. "It's the education that makes an Albanian. If we go away and lose our language and our culture, then we are no longer Albanian."

The others nodded in agreement. Having all been denied access to school under the Serbs and then driven into exile, the children did not see these as purely academic questions. Their Albanian identities were nearly destroyed not long ago.

"What makes a Serb a Serb?" I asked. Malesora repeated the question to herself, thinking hard.

"A *shkja*," she said, using the derogatory term for Serbs, "kills Albanians, that's how you know."

"Even if a *shkja* learned Albanian," Peter threw in, to make sure his point was clear, "he would still be a *shkja* at heart. He would still hate us." They slipped quickly, thoughtlessly, effortlessly to using the term *shkja* the same way their Serb peers used *shiptar* when talking about them.

The war had been over for five years. The children with whom I met on both sides were between eight and ten years old during the war itself. All of them wanted to be left alone, to live in peace. But few had found that peace that comes with forgiveness. The entire province still seemed to be in the mentality of war, hating the enemy, avoiding contact except to taunt or come to blows. The casual use of racist language was a symptom that reinforced the disease.

I had expected to find a society healing from the wounds of war, but the children I met were picking at one scab continuously. While, for the most part, their drawings expressed concern with the everyday problems of life—no more anguished pictures of murder and destruction as I'd seen in the ongoing conflicts in Burma and throughout East Africa. They drew pictures concerned with family and landscape—as some of the children in East Africa and Asia had as well. Politics and nationalism also found expression in the children's drawings in the Balkans. With the Albanians it was a longing for the future, as in one drawing that showed Kosovo as it was now—dirty, crime-ridden, crumbling—and Kosovo as it could be after independence—clean and prosperous with lovely homes and shining skyscrapers (Figure 21). Another drawing showed Kosovo at a crossroads, with war and drugs and collapse in all directions but one: the shining direction of independence (Figure 22). The Albanian children longed for an independent Kosovo the way the Serb children longed for Lazar to rise, messianic, and return them to glory. Their drawings held a nostalgic longing for the past they had only heard described, had only ever dreamed in stories.

These were palpable longings in old and young alike, two nations in one land; one nation longing for the past, one longing for the future, the present a neglected Dumpster, a barbed wire fence, a burning building. No one wanted the present, the here and now. The children's sense of who they are belonged to imaginary times, the distant past, the hoped for future.

When I asked the children in the Albanian village of Zahaq what independence meant, why they craved it so strongly, twelve-year-old Mark's answer summed up how the war still worked on their desires.

"Independence means no one can tell you what to do." The others agreed.

"When the Serbs were here they would not let us go to school," Nora explained. "They wanted to keep us undereducated so we would be easy to rule. With independence, they could not do that to us."

All the Albanian children believed they had a future if they could achieve independence, except for Karl, whose father survived the first forced evacuation of Zahaq only to be gunned down a few weeks later by a man in a yellow Mercedes. He said, "I probably won't live to be a grown-up."

Karl had trouble in school, had trouble sleeping and sitting still, I was told. He had problems beyond memories of the war and the loss of his father. His family was very poor. They struggled to survive in a devastated economy. Kosovo had always been the poorest province of Yugoslavia, but since the war and the post-war interim government, the poverty, everyone told me, had gotten worse. Unemployment was high. Alcoholism was high. Drug use was high. Depression was high.

Karl still struggled with the past and the present, reminding me that not all wounds heal with time, that even resourceful, intelligent children break under the strain of all the stress this world can heap onto them, that not everyone is as resilient as Valerie, the little Zahaq girl who told me that she survived because she must, because "life continues."

Nora gave Karl her advice. "You must live," she said.

She said she would like to tell this to all the children of war that I met in my travels, all the ones who were losing hope. "You must live," she said, and I thought of Keto in Lugufu Camp and of Nicholas on the Thai-Burma border, and of Charity from the Sudan and Paul in the Congo. All of these children were eloquent testimony to the idea that in a society at war, the smallest citizens carry a heavy burden, that of living, of continuing because one day peace will come, as it had, in its own way, to Kosovo. Because they were the future; because they would be the next ones to wage war or to make peace and maybe they could do things better than their parents had, if only they could take her advice and live.

As I mused on these grand thoughts, bigger than the room we were in, bigger than the girl who spoke them, thoughts that moved continents, thoughts that could end wars, that united the children of the world—in my mind—into one grand celebration of life and innocence and the resilience of man, Nora kept speaking. Her next words brought me back to reality, away from the dangerous, ridiculous, greeting-card idealizations I was drifting into.

"Except the Serbs," she said. "I would not give them that advice. To them I would say, go straight to Hell."

The others laughed, bright smiles nearly knocking me over. Karl's sad face broke into a smile. Even my translator, a bit of a nationalist himself, laughed.

Later in the day, Karl told me that he would be a soccer player when he grew up, a professional, and he would live in a free Kosovo. It felt good to hear him talking about dreams, expressing some vision of himself living to see adulthood. I had scrawled "Depressed? Post-Traumatic Stress Disorder?" in my notes next to Karl's name when we had first spoken, but these sorts of labels are perhaps too easily applied. As the psychiatrist Lynne Jones notes of children who survived the war in Bosnia, "these children might experience intrusive recollections of events, might have nightmares and difficulty concentrating in school," but when one takes into consideration where they live, what they've been through, and what their current conditions are, could these reactions be understandable responses to horrific events rather than symptoms of a psychiatric disorder?

In a depressed economy like Kosovo, in a society where everyone was the victim, how could one define what an abnormal reaction would be? In a time when the local high schools were turned into rape camps, when men were pulled from their families and shot then dumped in mass graves, who was sick in the head? Who was well? These were questions I was not qualified to answer, questions I think no one really is qualified to answer when it comes to the unique madness that is war. Karl's sadness was justified. Things around him were not improving.

"We still have many *shkja* here, you know," he said to me later, knowing that the subject of ethnicity interested me, knowing that I intended to talk to Serb children as well. A mischievous smirk spread across his face. "You could take some with you when you go. Take as many as you like."

If young people act as a mirror to society, these youngsters in Zahaq were an excellent mirror to Kosovo. They were full of hopes and humor, resilience and grit, and an unrelenting grip on history, so strong that it might keep them mired in their past hatreds forever.

In Babin Most, an isolated Serb enclave halfway between Mitrovniča and Pristina, the children were preparing to celebrate Vidov Dan, which Bujana, an eleven-year-old girl who was originally from Pristina, explained to me.

"We celebrate every year so we do not forget the Serbs who fought," she said.

"Fought where?" I asked, though I knew the answer.

"The Battle of Kosovo," she said. "In 1389." She then proceeded to tell me the story in much the same way the children in Lapjo Selo told it, though with perhaps a bit less emotion. She was a quieter girl, blond and engaged in the details of her drawing. She told the story as if she had told it a dozen times before, which no doubt, she had. It had the ring of a schoolroom report, which no doubt, it had once been for her. Her story lacked the spontaneity that the kids on the soccer field had when they told it, but the details were all there, just as they had been, the traitors, the martyrs, the names of the dead. Despite her dry telling, this story was her story.

To quote Slobodan Milošević's famous speech on the six hundreth anniversary of the battle, the speech that propelled him into leadership in the waning days of Yugoslavia: "Today, it is difficult to say what is the historical truth about the Battle of Kosovo and what is legend. Today this is no longer important. Oppressed by pain and filled with hope, the people used to re-member and to forget, as, after all, all people in the world do, and it was ashamed of treachery and glorified heroism. There-fore it is difficult to say today whether the Battle of Kosovo was a defeat or a victory for the Serbian people, whether thanks to it we fell into slavery or we survived in this slavery."

This was indeed how Bujana experienced the story of the battle, not as a collection of facts, which could be true or false but as a piece of herself and her people, how they suffer and they survive.

Oorus, her cousin who was two years older than Bujana, had more pressing concerns than the old stories. "Five years ago," he said, "we could walk around without being afraid. But during the war, Albanians kidnapped my grandmother and grandfa-ther in Pristina. We fled to Montenegro for a year and then came back here. In this village we don't have Serbian televi-sion or movies like other kids. We don't have any freedom to go anywhere. And no money either. I want it to go back to how it was five years ago. People should go back to their old apart-ments and jobs."

"Freedom to move would make me happy," Bujana said.

Two young children were visiting from Belgrade. They sat in silence while she spoke, doodling with the crayons I gave out.

They did not have the problems of their cousins in Kosovo. Their cousins were country bumpkins; somewhat backwards, isolated hicks, but family nonetheless. And better than the *shiptar*, they made clear. My translator spoke their words faithfully, though he himself was *shiptar*, part of a multi-ethnic human rights group in Pristina. He had known the families in this Serb enclave for years and many trusted him despite his ethnic background. The atmosphere in the room cooled a bit as the two kids from Belgrade spoke. Perhaps they did not realize an Albanian was in their midst. How could they have known? Everyone's accent in this part of the country sounded funny to them, and they were far less attuned to the ethnic signs than their peers who lived and died the secret code of whom you belonged to, who your people were.

It was strange for the visitors to talk about these things, they said, because in Belgrade, Serbs were in control. They could go where they pleased; they had access to movies and television and all the things of city life that Oorus and Bujana lost after the war in 1999. Their drawings were of houses and cars. Bujana also drew a house, but hers was surrounded by thick black coils making a tight outline. Trees, flowers, roads, everything was on the outside of the coils (Figure 23).

"Barbed wire," she said. "Because we are trapped here. If things don't get better, we will leave Kosovo for Belgrade. Most of the young people leave."

"Do you want to leave Kosovo?" I asked her.

"I want things to go back to the way they were five years ago."

"Do you blame the Albanians for what happened?" I asked her.

"Politicians cause the problems," Oorus said before Bujana spoke.

I would only realize when I spent that night in Babin Most without my translator how much his presence changed the children's attitudes. Even though they had known him for years, none of them trusted him. He later acknowledged this, but told me it was the only way things would improve. He would just have to keep going out there.

He left after lunch, and Oorus and I sat in his TV room. He poured me a glass of *slivovitz*, a potent homemade plum brandy. I marveled at his access to it, as he was only thirteen years old, but he poured himself juice. There was plenty of time in life in the Balkans to discover alcohol. He was in no particular rush. It would be there for him when he grew up, he joked. I might have preferred juice. Oorus's mother had insisted I eat lunch with them, a real traditional Serbian meal, with cabbage and beef and cheese, pastries, a tomato and cucumber salad, and several glasses of *slivovitz*. I was a little drunk. Oorus wanted to practice his English with me. With *slivovitz* coursing through my body, I felt I could practically speak Serbian.

"You like Kosovo?" he asked.

"Yes I do."

"You stay Pristina?"

"Yes," I said. "I am staying in Pristina."

"Pristina is not good now."

"Not good?"

"Many *shiptar*," he said, using the pejorative he had avoided while my translator was present.

"You don't like Albanians?"

"All terrorists," he said.

"Terrorists?"

"I hating them," he said and ended the conversation by insisting we watch his favorite video, a fuzzy tape of *Children of the Corn 3*. He did not want to talk about the past any more.

A few hours later, he took me to the school to watch the traditional folk dancing group practice. I helped him tie a braided belt that makes up part of the costume for the traditional *kolo* dance. All the children I had met that day were part of this dance group, and they were glad to have a visitor. Bujana danced with particular enthusiasm. Oorus, it seemed, was a bit embarrassed by the pageantry, but it was part of the Vidov Dan celebration every year. A group of older teens played the music and the children moved and swayed in circles, kicking and bowing and linking arms. It was a remarkable dance, one they had all practiced countless times.

"Folk dancing," Oorus's mother said to me, pointing, exhausting her English with that. She called Bujana over to us from the dancing circle. The teacher looked annoyed but let her go, deferring to their guest's need for someone who spoke English, even if it was his eager eleven-year-old dancer. Oorus' mother spoke to her for a moment and Bujana translated for me.

"She say folklore is good to have. It make children good." Bujana didn't have the English skills to clarify what she meant, and we couldn't find anyone else in the remote village who could ease communication, so I was left to wonder what she wanted to tell me about the folk dancing. Folk dancing kept children out of trouble? Folk dancing gave children a positive sense of their national identity? Folk dancing kept them fit?

Oorus's mother tried to tell me that she could not say these things in front of my translator earlier, that he could not translate them. Then again, neither could she. Frustrated, she let Bujana's translation abide.

I got to thinking about Oorus. With my translator present, he gave lip service to the notion that people, no matter their ethnicity, were good and that the politicians caused the problems in Kosovo. At the time, that comment struck me as thoughtful and accurate. It gave me hope that reconciliation might be possible. But his comment when we were alone, which was made offhand and which, I can only assume, was an uncensored expression of his opinion, dashed that hope. He knew enough of the world to give lip service to rhetoric of forgiveness, but in his heart, he carried all the old prejudices.

Albanian children I met said much the same things. When I would ask if it was possible for Serbs to be good, they would say, "Of course, not all people are bad." But when pressed, they would tell myriad tales of their Serb neighbors failing them, informing on them to the police, looting their homes after they fled, throwing stones at them, joining the paramilitaries that raped and killed Albanians. By the end of the discussion they would tell me, as Nora in Zahaq said, "We can forgive them if they are punished for their crimes, but they should not come back here. There is nothing for them here."

I wondered how reconciliation was ever going to be possible when neither side would give any ground, when the children, like the politicians, gave lip service to forgiveness and cooperation but never showed any signs they meant it. Of course, this is not the children's fault. They have no exposure to youth from other ethnic groups. Serb and Albanian children live in complete isolation from each other. How much difference, I dreamed, would a mixed ethnicity soccer game make?

I talked to one boy, Milos, whose family had fled to the Gracanica Monastery when their home in Obilic was burned to the ground. Three children, the mother and father, and the grandmother lived in a small outbuilding on the grounds of the five-hundred-year-old cloister. The father complained that the Albanians were a people without culture as his son stood beside him, saying little. His father went on and on about the politicians screwing the people. I asked Milos about his school in Obilic. His father answered for him.

"It was like a prison," he said. Milos nodded. "Bars on the windows. The children couldn't even go outside to play, because the Albanian kids provoke them. The teachers can't control the kids. It's like they're drugged."

"Did you ever have problems with the Albanian kids?" I asked Milos, trying to get him to speak instead of his father.

"Yes," he said quietly.

"Tell them," the father said. Niko, my translator, rolled his eyes very subtly to me. He was trying to get the father to back down and let Milos speak naturally, but the father was not having it. Milos did not want to talk to us and we decided to let him be, but the father insisted. I felt very awkward, as if I were forcing the child to speak against his will. "Tell them," he said again.

"When we walked around at school, the Albanian kids came over and provoked us with cursing and swearing. They would throw stones. There were only thirty-four kids in my school, and we were right next to the Albanian school."

"What would you do when they provoked you?" his father asked.

"I'd do the same as they did," he said. His father patted him on the head. This was not a "turn-the-other-cheek" kind of family. Nearby, a cat had given birth to litter of kittens. The kittens kept slinking into the small building and the grandmother, a powerfully built stoop-backed old babushka, emerged from the doorway shouting and cursing at the kittens, tossing them through the air by their hind legs. We all stopped a moment to watch the kittens fly, one by one, across the lawn.

"The Albanians have been like this since '99," Milos' father said, and for a moment I thought he was saying that they Albanians had been like the kittens, flying through the air, screeching. But he was talking about the abuse they heaped on the Serbs.

I remembered images from the news in 1999 of Pristina after the Serb pullout. Chaos ruled. Celebrations turned easily to riots. People felt exalted, untouchable. The Kosovar Albanians had never had their own homeland and it seemed, in 1999, that they might get it. For a time, law and order broke down completely. The Kosovo Liberation Army became the de facto law, acting more like a mafia than like the police.

In 2005 Kosovo was still technically a part of Serbia, though administered by the United Nations Mission in Kosovo (UNMIK). UNMIK created a police force shortly after the fall of the Serb government, as the KLA rule deteriorated into violence and thuggery that had many Albanians as frightened as they had been under Milošević. But six years later, the international police force still kept law and order and the UN still ruled the province. Albanians were frustrated. They wanted independence and felt the Serbs were holding them back. Milos' father wanted Kosovo to remain part of Serbia. He was a farmer and was not about to leave his land. He was proud to tell me that he had no debt, no debt at all. With massive unemployment in Kosovo, especially for Serbs, this was no small feat. He wanted to pass something on to his son. He had survived the war and the post-war politics, only to lose his home to a mob. His rage was palpable and had become the only thing he had to pass to his son. Inherited anger can be the most venomous. Milos smirked when his father patted him on the head, proud of how he had thrown rocks and curses right back at the Albanians, proud of his own little war.

I left Gracanica Monastary feeling more despondent than

ever. Children observe, they take in the adult world moving around above them. From Southeast Asia through Africa and into this broken corner of Europe, I marveled at the way the children watched everything and learned to navigate their war zones with deftness and, often, grace. Those who survive were usually those who were most observant, most engaged. When the adult world lumbers above like blind elephants, the children, the little mice, learn to scurry from underfoot, learn to march behind the elephants, following in their footsteps. It is a matter of survival.

In Kosovo, where I expected to find youth in the process of reconciliation, the adults were modeling the most dangerous behavior for their children, the behavior that led only to war. The children observed the hatred around them, observed the divided society, and adapted to its norms. I had seen children in other war zones choose forgiveness and peace, and knew that the Serb and Albanian children I met were capable of choosing it too, even if no adults around them modeled it. How else to explain child soldiers like Musa, Xavier, and Paul in the eastern Congo? They were not blessed with positive role models either.

Why then had I yet to find one child in Kosovo who did not, in some way, demonstrate that the ethnic war was not still raging in their hearts? Why did they all hold onto these ethnic hatreds that had caused them and their families nothing but pain and loss?

"You must understand the history," my friend Alex told me.

He was a medical student in Pristina and had been a refugee in Germany during the worst of the war and the aftermath. He was Albanian, but by no means a fanatic nationalist, though he had some involvement in the independence movement as a teenager in the Milošević days. He was more concerned with his American girlfriend, passing his first year medical exams, and finding a rare Slipknot album than with ethnic conflict and politics.

His Serb neighbor had been a paramilitary. His own apartment had been turned into a clubhouse for Serb death squads after his family fled. A bit more cosmopolitan than the youths I was meeting in the village, Alex had no special resentment for Serbs as a group, though he did not really believe coexistence was possible without independence. Many of his friends were involved in multiethnic peace organizations (such as my translator in Babin Most). I feared that even he was about to bring up the Battle of Kosovo.

"I know, I know, I know . . . *the history*," I groaned, contemplating throwing myself out the window rather than hear the story one more time. "1389. Kosovo Polje." I took a long swig of my beer.

"Sort of," he said.

"What?" I asked.

"There is never peace in the Balkans," he said, laughing. "We just can't help killing each other."

"Come on," I said.

"No, really. There was the Ottoman takeover. And the Austro-Hungarian war, uprisings right and left. World War I started here. And then World War II was fought here. Under communism, there wasn't so much violence. But in the eighties there were marches and riots in Pristina. Then Bosnia, then the war here. There is never peace for long. Some people say, 'It's been five years, we're due.' No one really thinks there will be peace."

At talks in the summer of 2006 about the future status of Kosovo—whether it would remain a part of Serbia, though granted greater autonomy, or whether it would gain independence—as 90 percent of its inhabitants desire—no hope of a peaceful settlement was in sight.

"Belgrade was willing to give everything but independence and Pristina wanted nothing but independence," the UN special envoy at the talks said. As a BBC article reported, the Serbian prime minister refused to go to a joint lunch with the Kosovo delegation, and neither side offered a handshake when the talks began.

The adults, the ones in power, the ones who made the speeches, the ones that made the wars that rocked the children's lives, modeled no behavior that could suggest an alternate view of the situation: the Albanians and the Serbs would not, could not get along.

The children have television, access to magazines and newspapers, or, if not, they have the rumor mill or the conversations going on among their parents and the other grown-ups around them. They learn about these events; they watch and listen and learn, and in Kosovo all they see is the madness of ethnic hatred and failed diplomacy. They do not live in a world where talking things out leads anywhere. They live in a world where the other wants to control them, to get rid of them, to exterminate them. The more engaged the child, the more aware they would be of the problem and the more helpless they would feel to prevent it, to change it. *If the leaders couldn't even sit down with each other. . . .*

Could that explain it? The children held onto these opinions because they were girding themselves for another war, another war that seems more and more likely? They acted like the grown-ups around them not because they inherited bigotry passively, the mindless recipients of their parents' worldview, but because bigotry was more practical.

They *chose* bigotry.

After all, who would want to reach out to the other side if that would label you a collaborator when the violence erupted again? Staying in a mental state of war with the other side prepared these children for the inevitable war. The thought did not sit easily with me: bigotry as a survival strategy, each child's internal, quite personal war without end.

Christof jogged over from the soccer field where the oth-ers were playing. I sat in the shade of a newly built pavilion petting the stray dog that had found its way to us on the slopes of Mount Igman, an hour outside of Sarajevo, Bosnia. Christof tried to practice his schoolroom English with me, the look of concentration, the search for those vocabulary words the teacher always blathered on about.

"Here, for you," he said and stuck out his arm, dropping a rusty shell casing into the palm of my hand. He'd found it in the dirt on the soccer field. It was the third one he'd found that day, and it had become something of a project for him, picking up rusty shell casings, grim reminders that the place this youth group had chosen for its weeklong summer excursion had once been the site of bloody fighting between Muslim and Serb forces. The casing was as long as my index finger.

No one was allowed to wander off into the woods. Ten years after the war in Bosnia ended, land mines still hid in the forests, regularly exploding unsuspecting deer. The dog at my feet looked up at Christof with a panting smile. It was hot as hell on that mountain and the dog, a stray who had wandered around the mountain and had become a somewhat unwelcome fixture around the guest house, rested happily in the shade of the pavilion. Despite his size and rather tough look—he was a Rottweiler-German shepherd mix—he had a deep love of people and warmed up to the children immediately, following us around from activity to activity. I marveled at his survival, wandering as a stray through these deadly woods on this deadly mountain.

That he had not set off a land mine was a miracle. He had a wound on his mouth that made him look ferocious—he was at least fifty pounds—but he clearly loved people and loved getting scratched behind the ears and on his amazingly soft belly, which I gladly invested a good deal of time doing. Christof did not join me in petting the dog, but he looked at me puzzled as I did. He watched me looking at the shell casing he had given me, scratching the mutt, and generally pondering the sordid history of this mountain.

"Ugly dog," he shouted suddenly and mimed kicking the mutt in the stomach.

It was my turn to look puzzled now. Christof bolted off again to play soccer without giving any explanation for his outburst. I would join him soon, but I needed another moment alone with the dog that we had named Prijatelj, which means "friend" in Bosnian. He was a mixed breed dog, which made him unpopular in a country where the term "ethnic cleansing" had been coined less than a decade earlier.

I watched Christof hit his stride—he was a graceful thirteen-year-old who took great pleasure in a well-placed slide-tackle or an elegant header. It was hard to believe, watching him leap off another boy's shoulders to head the ball toward the goal (he missed and laughed about it to convulsions) that he could be filled with so much hate toward this beat-up dog. I had come from Kosovo a few days earlier and had grown used to bigotry among people, expected it even. But hatred toward this one dog? I couldn't understand it.

I had stopped him and his brother more than once from throwing rocks at Prijatelj. Christof was blond with piercing blue eyes, just becoming aware that girls liked how he looked though he was still uncomfortable with himself. He picked on other kids, the smaller ones, the weaker ones, or the ugly ones.

But when the adults around scolded him, he was not only quick to apologize to the other kids, he showed them real kindness. He included them in his games or even, in the case of a tiny Jewish girl named Sofya whom he had been mercilessly picking on for days, carried her on his back during a hike when she grew tired. He struggled to figure out his place, the kind of young man he wanted to be. In that struggle, though, he had decided where this mutt stood. He hated the dog with unapologetic venom. Even in the presence of the adults (there were three Americans, including myself, and two older women from Sarajevo who worked with the youth group year round), Christof's hostility toward the dog did not abate.

Christof, I learned, was a mixed breed too—half Serb and half Croat—and during the siege of Sarajevo, no one trusted his family, though they suffered the deprivations and shelling like everyone else. They lived in Grbavica, the one neighborhood in the city controlled by the Serbs. The neighborhood was on the front lines and was as dangerous a place as any. Bosnian and Serb snipers faced off and fired mortars at each other, killing anyone unfortunate enough to be inside the targets or in the range of the shrapnel. A fragment the size of a roll of lifesavers could take off your head. Death came easily in Grbavica, just as it did in the rest of the city. When the siege ended, however, and the Serb lines moved back and the city became one again, Christof's family ran into other problems.

Neighbors scrawled offensive and threatening graffiti on their door; the kids were picked on in school, bullied. They were half Serb; they'd lived in the Serb-controlled area. They were the enemy. SFOR—the NATO peacekeeping force in Bosnia—had to intervene to protect the family from harassment and threats of violence. The two boys, Christof and his little brother Peter, who was just a baby at the time, found it very difficult to have foreign soldiers protecting their family. It somehow seemed to amplify their guilt and send a message, in that nasty schoolyard way, that they couldn't take care of themselves. Something along the lines of crying to the teacher because of a bully, though in this case the bullies were myriad and the "teacher" carried M16s.

The mockery in school did not stop. NATO doesn't concern itself with child's play. Christof learned to bully from the bullies, learned to shoot an insult like a sniper's bullet because the insults were shot at him like mortars. No one trusted them; no one wanted them in their neighborhood. He and his family belonged to some other group, the enemy.

This was Christof's first and only summer in the multiethnic youth group on Mount Igman. All five of us adults were determined to help him see things another way, to help him feel like part of the group, and I thought, perhaps foolishly, that this big good-hearted mutt could help if Christof would just stop throwing stones at it.

The multiethnic youth group with whom I was visiting on Mount Igman, run by the Jewish community in Sarajevo, began informally during the siege. As the former communist nation of Yugoslavia began to break apart in 1991, the Jewish community saw unsettling signs of rising nationalism—new flags and slogans, angry rhetoric, guns everywhere—signs that brought back painful memories of the Holocaust.

Before World War II, Jews made up 13 percent of Sarajevo's population. By the time the war was over, they made up less than 4 percent Their numbers dwindled, as they did for all of Europe's Jews at the hands of Nazi death squads or in concentration camps. Still others escaped to join the partisans and fight the German occupation, dying in combat to liberate their homeland from foreign invaders. Just down the hill from the soccer field stood a monument to the fallen partisans who liberated Yugoslavia from the fascist regime. Many Muslims and Christians in Bosnia hid their Jewish neighbors from the SS or the Ustashe (the Croatian branch of the Nazi party, essentially) at great peril to themselves. The war ended, and Marshal Tito took control of Yugoslavia, forging arguably one the most successful communist states in the post-war era.

Under Tito's rule, ethnic tensions were suppressed and Jews were treated like any other member of the state. Yugoslavia tried to stand as an example of cosmopolitanism among the communist nations, and Sarajevo was the prime example. They touted their ethnic and economic harmony, wore it like a badge. They were the intersection of East and West, Muslim, Orthodox Christian, Catholic, and Jew. The architecture in the city was a rich assortment of styles and eras. Sarajevo hosted the 1984 Winter Olympic games, building a lovely Olympic Village, a new stadium and hotels, and a mammoth Alpine Jump on Mount Igman. Muslim, Jews, Catholics, Serbian Orthodox— all called themselves Yugoslavian.

This picture of harmony masked the nationalism brewing underneath the surface. Throughout the seventies and eighties, Serbian nationalism began to grow. The Serbian Orthodox Church aligned itself with Serbian nationalist politicians, largely due to the crisis in Kosovo. Serbs held a majority of government posts. At the same time, the Islamic revolutions around the globe—most notably in Iran—began to inspire Muslims to claim their own identity as a national group rather than a private religion as they had been defined. The Croats, Slovenes, and Serbs all claimed nationality, and the Muslims wanted the same. Agitation to that end met with hostility from the Serb majority in the Communist Party and the argument that would eventually lead to the destruction of Yugoslavia and the start of the Bosnian war began to roil with accusations of repression and religious fanaticism.

But the Jewish community who remained in Bosnia—some two thousand of them—never forgot who they were, though they generally tried to stay below the radar of nationalist debates. They tended the old cemetery in the hills, where the graves dated back to the sixteenth century. They maintained the old Sephardic synagogue in the Turkish quarter; they understood how easily a people could be lost, despite the illusion of harmony and prosperity, and how dangerous the revival of ethnic nationalism could be. They also remembered the great kindness many of their neighbors had shown them during the Nazi time. Perhaps more than any group in Yugoslavia, they remembered the sting of war, which put them in an historically unique position in the early nineties: the ethnic conflict that erupted was not about the Jews.

At their annual meeting in Belgrade in June 1991, arguments between the Jews of Serbia and Croatia broke out over the question of independence. That month, Slovenia and Croatia had declared their independence from Yugoslavia—something Belgrade opposed because 12 percent of Croatia's population was Serbian. The arguments between the Jews living in Serbia and the Jews from Croatia were fierce. There were loyalties among the Jews, of course, but these Jews were also members of nations and not completely immune to the patriotic fervor of the time. The Croatian Jews argued for the necessity of an independent Croatia, while the Serbian Jews argued with equal vigor for the maintenance of a state led by Belgrade.

The Bosnian Jews found themselves isolated from the other Jewish communities, who spent much of the meeting arguing with each other. Loyalties divided between the family of Yugo-slavian Jewry (who had met together every year since the end of World War II) and the national identity of the Jewish mem-bers. The tensions at the meeting made it clear that greater violence was coming. If the Jewish communities fell to bicker-ing among themselves, what would happen between the other ethnic groups?

Jacob Finci, the head of La Benevolencija, the humanitar-ian arm of the Bosnian Jewish community at the time, recalls a visit to the beach in Croatia that summer when, in the parking lot, he experienced of shudder of terror. A group of men draped a Croatian flag over their car and stood around it drinking and singing patriotic songs. What struck Jacob was how much the nationalist flag of Croatia resembled the flag he had seen as young boy, fifty years earlier: the flag of the Ustashe, the Croa-tian Fascists. It was then that he was certain there would be no more peace in the Yugoslavia.

As independence movements stirred throughout the Bal-kans, tensions were highest in his home country of Bosnia, the most ethnically mixed of the former republics of Yugosla-via. Radovan Karadzic, the leader of the Bosnian Serbs and a trained psychiatrist (and, oddly, an amateur poet of mediocre ability), stated that if Bosnia tried to declare its independence, they would be on a "highway of hell."

When the Bosnian parliament overwhelmingly passed a ref-erendum on independence in April 1992, the Serb MPs boy-cotted the vote. Soon afterwards, both sides erected barricades around the city.

On April 6, 1992, during an independence rally, Serb forces and Bosnian police clashed, starting the civil war. True to his threat, Radovan Karadzic and the head of the Bosnian army, Ratko Mladic (both currently under indictment for crimes against humanity), did indeed turn much of Bosnia into a hell for the Muslims: mass rapes and massacres spread throughout the country. Under the banner of Orthodox Christian zeal and patriotism, Serbian nationalists aimed to "cleanse" Bosnia of its Muslim population. They forced Muslims from their homes, stole their identity papers, and destroyed their mosques and cultural sites. While Muslim forces were also guilty of human rights abuses, little evidence suggests that targeting civilians was a matter of policy as it was for the Serb army. The Serbs hoped to hold Bosnia as a part of Slobodan Milošević's aspirations toward a "Greater Serbia."

Tanks and artillery surrounded Sarajevo, which sits in a valley, and sealed the roads in and out. For the next three years, cut off from the outside world, the city was bombarded and starved while the Bosnian army, made up of local conscripts and police, tried to keep the city from falling.

When they returned to the city from that 1991 meeting, less than a year before the siege began, the leaders of Sarajevo's Jewish community, Jacob Finci and Ivan Cersenjes, started to organize. They gathered the community's doctors together and talked about what they would need. The Jewish Joint Distribution Committee, based in New York, helped them prepare, sending them stockpiles of medicine and bandages—and, at the doctors' grimly pragmatic urging, body bags.

On the first night of intense shelling, late in April 1992, Ivan Cersenjes returned to the community center to work. To his surprise, he counted about sixty people sleeping on the floors and benches as he walked around. He didn't know who they were. The building was open to the public, so he assumed they were frightened people from the neighborhood who did not know where else to go. From that day forward, the Jewish community never closed its doors to anyone. Within days of the outbreak of war, Jacob Finci had called his counterparts in Zagreb and Belgrade, who, despite any nationalist sentiments they might have had, were eager to help. Throughout the conflict, Finci used all his connections on all sides, and, with skillful negotiation, the Jews were able to keep their supply lines in and out of the city running.

But it was not just their neutrality and their influential friends that helped them. According to Yechiel Bar-Chaim, a Joint Distribution Committee field worker in the region during the conflict, "an awareness of the Holocaust—of what had already happened to the Jews of this region—was one of the elements that everyone had in mind." Empathy with the Jews, Bar-Chaim points out, was nearly universal. In the face of their own suffering, each ethnic group identified with the Jews. Parallel to this empathy, which certainly existed on the Serb side, who saw their entire history as a series of wrongs committed against them, there may also have been a fear of international intervention if Europe or the United States suspected a second genocide attempt against European Jews. Empathy and pragmatism combined to allow the Jewish community to remain supplied throughout the war, ensuring not only that their own would be fed, which was a first priority, but that they could feed as many of their neighbors as possible too.

They opened three pharmacies—"the best pharmacy in town," according to Finci—created a clinic in the community center, and arranged for doctors to make home visits to patients who could not make it out of the house. A psychologist who worked with the community immediately after the war told me that, if not for their arrangements, many more residents of Sarajevo would have died.

Muslim and Croat nurses, doctors, and volunteers dashed through the streets, risking their lives under the sights of snipers and under the mortar blasts in service of the Jewish community's aid operations. According to one report, of the 60 La Benevolencija employees during the war, 19 were Jews, 19 were Muslims, 13 were Serbs, and 9 were Croats. The Jewish community worked with everyone. There seemed to be no doubt among any community members that their doors should be open for all.

They even arranged for bus convoys to evacuate "members of the community" from the city. They stretched the definition of who counted as a member of the community as far as it could go. Of the three convoys that left Sarajevo and took refugees to Croatia, more than half the evacuees—over a thousand people—were non-Jews.

Many members of the Jewish community chose to stay. Sarajevo was their home. There were elderly community members who could not leave, who were housebound or confined to the nursing homes. These were their people too. The leadership also stayed. They were just as determined as members of the army and government not to let Sarajevo fall. How could they simply abandon their friends and neighbors?

The siege turned Sarajevo into a nightmare. Mortars and grenades crashed into all parts of town indiscriminately. Machine gun fire tore the streets apart. The city lost power. It lost water. People burned furniture to stay warm. Snipers shot women and children running with buckets full of water on their way home. The Olympic stadium, once a symbol of Sarajevo's unity and prosperity, became a different kind of symbol. It was transformed into a graveyard. Bodies filled the field.

Jaca, who was thirteen during the war, walked with me toward Bulevar Mese Selimovica, the street that was known as Sniper Alley, which she told me I must see to understand how the people of Sarajevo lived. When we arrived, she warned me, she might break into a run, "just out of habit," she laughed. "Try to catch up with me if I do. You'll look silly if people see a girl running away from you. Many people still duck and sprint when they reach this road. The war, you know," she smiled. "It made us all insane."

Up above us on the hills, I could see the Jewish Cemetery. Snipers had a straight shot to where we stood, to this entire stretch of road. They hid behind the gravestones, which provided excellent cover. The hid behind the graves and they rained bullets on the city. Rumor had it that visitors who were so inclined could pay for the privilege of taking a few shots into Sarajevo for themselves. During the siege, anyone who walked or drove along this road risked taking a bullet from one of these snipers who did not distinguish between civilian and combatant. Life was cheap. Cars would race at dangerous speeds, swerving up and down Sniper Alley from the airport trying to make themselves harder targets. The city erected metal barricades to protect the sidewalk from sniper fire, but the Serb guns quickly turned those to Swiss cheese. Jaca too looked up at the cemetery and sighed.

Now twenty-six, a writer and translator, she is a vivacious woman, a stunningly beautiful brunette with a dry sense of humor and a lot of plans for the future. She's written a children's book and plans to write others. She has a great love for children's literature, perhaps because the war that cost her her father and her best friend, many of her friends, also cost her her childhood.

"I never thought about ethnicity before the war," she told me. "My grandmother was Jewish, my father was Muslim, my mother was Catholic. We were a little of everything. When the war started, my friends began to ask me 'What are you?' I said I didn't know. They told me I was a Muslim, so I came home and told my father I was a Muslim. The next day, my Catholic friends said, 'No, you're no Muslim. You're a Catholic like us.' So I went home and told my father I was a Catholic. But it didn't matter. We were all in the same situation in Sarajevo. Catholic, Jew, Serb, or Muslim. We were all in the same kind of hell.

"For example, on my eleventh birthday, I was walking to school. I lived near the Holiday Inn and had to cross this street here, just up ahead." She pointed toward the infamous yellow building, a branch of the global hotel chain, which, during the war, housed most of the international press corps, though the rooms facing the street were uninhabitable due to sniper and mortar fire. I noticed with some degree of horror that the building was surrounded by tall apartment buildings that housed thousands of Sarajevans during the siege, many of whom could not simply take rooms away from the street as the journalists could. Their homes were easy targets, and many had to be abandoned for safer ground. Though nowhere in the city was safe. Death was always above it in the hills that ringed Sarajevo, sealing it.

Ten years since the siege ended and Jaca began to grow tearful as she told me her story, a story about walking to school.

"I was walking with my friends when a bullet hit the pavement in front of me. I turned to run backwards but a bullet hit there too, so I lay down on the road next to my friend, like we had been taught. My friend, though, had been shot in the head. I saw him bleeding, dead. It was very sad because I had a crush on this boy; he was my best friend. I thought I would have profound thoughts when I was about to die, but all I could think was that it was good this day was my birthday. My parents would save money on engraving the headstone. They would only need to put one date on it and they would save money on throwing a small party for me that night.

"I lay there for hours that day. The sniper would shoot near me sometimes to tell me he was still watching. The UN tank came and escorted me and the other children to safety, and we left my little friend's body on the street there. Others would come get him to bury him—we could not do it. We were just schoolchildren, you know."

By the time she finished, she was crying, thinking about her ruined birthday and her lost friend. "Everyone has stories," she said. "When we hear thunder, everyone in the city ducks for cover. I hid in my bathtub during the celebration for the first day of the Sarajevo film festival one year, before I realized it was only fireworks."

"There were a lot of terrible times," said Dada Pappo, a stylish woman in her forties who has worked with La Benevolencija since the war. She sat in her beautiful apartment near the center of town and pointed out to me all the places where bullets and mortars hit. As she spoke, we drank her homemade liquor, *travaritza*. She gestured with elegant fingers that held a powerful Bosnian cigarette. Turkish delight was laid on the table in front of me. "Even in the war, we tried to live well, though this was not always possible," she said. "A mortar destroyed the apartment above mine. If the neighbors had not been staying with me at the time, they would have been killed."

She talked with one of her neighbor's children, who had been a young boy at the time of the war, about how they used to gather together to sing and play the guitar. They laughed as they shared memories of their time under siege.

"You remember the night when they were shelling upstairs we had a fashion show down here?" he asked her. Dada laughed and talked about the outfits they put on and how they strutted and wore crazy hats as if they were on the runway in Milan or Paris, all by the light of flickering candles and the rumble of crashing artillery.

"You know," said Dada, "it was during the war, those were some of the best times too. We laughed and played music and we all came together. Life was very full in those years, with the good and the bad."

I sipped her liquor and tried to broach the subject of ethnicity. I was not sure how to ask. She had worked with the Jewish community for years, all through the siege of Sarajevo, risking sniper fire and mortar blasts to cross the river to get to work.

"Are you Jewish?" I blurted, somewhat tactlessly. Her homemade liquor was strong.

"Me?" she smiled. "No. I am not very religious, though I was raised a Muslim."

Dada, along with a Jewish community member named Giselle, was at the forefront of creating the children's group.

During the siege, the Jewish Community started a Sunday School for the children of community members to teach them about their religion. They watched videos and heard lectures about Judaism. I imagine, with so many of the community members fleeing the city, the motivation for these lessons was one of preservation. While religious observation was never a high priority for the rather secular Jews of Sarajevo, they feared, perhaps, a loss of the last cultural ties to Judaism in the Balkans. They were cut off from their cemetery in the hills on the front lines. To this day, there are bullet holes in even the oldest of the graves, dating back to when Ladino-speaking Jews arrived in Sarajevo fleeing the Inquisition. Land mines littered the rows between the graves. For years, one could not come up here safely. Cut off from their dead, the Jews used the Sunday School to keep their history and their culture alive as best they could.

Early on, the school began attracting non-Jewish children. Children would come to the program and ask if they could bring their non-Jewish friends. Activities were limited during the war, and this program helped everyone take their minds off the terrible events around them. The community had clowns perform and threw parties in addition to the religious classes. During the Jewish High Holidays, non-Jewish children from the club were invited to visit temple services. The club members also visited major Orthodox, Catholic, and Muslim services. By understanding and experiencing each other's faith, the hope was that future ethnic conflict could be prevented. The multi-religious program for children continues today.

After the war, with UN troops patrolling the city, the leaders of the Jewish community decided to continue reaching out to the children. Their children's program, called Club Friends, has met every Sunday since the end of the war. The members come from all ethnicities.

Primarily, the club was a venue in which to address the traumatic experiences that every child had during the war. Every summer, the Jewish community took the club on a trip into the mountains outside the city for a week. Giselle and Dada, along with other leaders of the community, hoped to create a safe space for children in which fun would flourish and the troubles of city life—where poverty and crime are still constant worries—would be left behind. It was a place to find childhood again. To this day, the scars of the war are visible on buildings in Sarajevo—many burnt out shells of structures stand in the city center. On Mount Igman, the community tries to put all that behind them and look to the future.

"At first," explained one mother, " the children didn't know how to play. They had not been able to play outside for so long. They were very nervous." Even now, years later, mementos of the war still interrupt play. The shell casing Christof dropped into my palm spoke of past horrors, horrors no one could fully forget. It was these memories, the memories of a time when everyone was forced to identify with one group or another in a way they never had before, that the club hoped to erase, but, looking at Christof, I could see much damage had been done. His loathing for our adopted mascot, the mutt we called Prijatelj, was a loathing of ethnicity, his own awareness of himself clashing with the identity imposed on him.

I spent a lot of time over the four days on the mountain with Christof, kicking the ball around or walking without saying much. I tried to reinforce in him a sense of the validity of our dog's existence, that he was no better or worse than other pure-bred dogs (there were two beloved German Shepherds at the guest house), but had everything to do with how loving and smart a dog he was and how amazingly strong he was because he survived. Christof listened, and at times when he thought no one was watching, would pet the dog, once even bringing him water, but there was no Aha! moment; no great breakthrough in his defensive shell. To the last day on the mountain, he still bullied the mutt, still shouted at him and called him names, his face red with anger that this ugly monster, this mixed breed, should live. Over the course of the next year, I learned, Christof and his little brother stopped coming to the club events on Sundays, and neither returned to Mount Igman the next summer. I like to think my experiment in historically conscious psychosocial animal husbandry had some effect on his thinking, but I doubt it. His anger ran deep, deeper than a dog could cure.

Luckily, his anger was not the norm. Most of the children in the group got along quite well. There were few conflicts among the kids in the club. One boy told one of the American women I was with, a psychologist who had been coming to the group since the war ended, that the kids did not want to "mess things up the way their parents did." She too tried desperately to show the children the worth of the mixed breed dog. In our minds, he became Bosnia. He was a survivor, beaten, but living still, and determined. We wanted the children to see that being purebred was not the most important thing, not so important at all. It was all about who you were inside. The dog was an uphill battle, a stretched metaphor perhaps, for which the kids had no time. They had real identity battles to fight and at the moment, real fun to have. No time to play with a big ugly dog they didn't like, even if that "meant something" to the visiting psychologist types. The business of childhood continued.

The adults worked hard to create a safe space for young people, where the ethnic tensions that surrounded them did not have to exist. The children too were committed to this idea, for the most part. They wanted the past to be the past and did not dwell on it, did not show any eagerness to talk about these issues—it was talking about ethnicity that caused all the problems in their minds to begin with. History was best left below—on Mount Igman at least, soccer or arts and crafts were the more pressing concerns. The dog was not popular, perhaps, because he was a reminder.

I played soccer most days and watched Christof passing the ball eagerly to his Serb, Jewish, Croat, and Muslim teammates without hesitation—well, with a bit of hesitation because he could be a showboat in the game and wanted a bit of the glory for himself. As they all played together on the field strewn with shell casings, I took a short hike to the ruins of Hotel Igman and the abandoned Alpine Slide from the Olympic games, now riddled with bullet holes. In spite of the history right in front of me, the physical evidence that games are no match for destruction, I got the sense that as long as the children kept playing soccer, there was hope.

The children certainly had their resentments and their prejudices and their fears. Their futures were uncertain—when we left the mountain, economic hardship awaited every one of them. War criminals were still at large. Neither Ratko Mladic—the general responsible for the massacre of nearly 7,000 civilians at Srebrenica—nor Radovan Karadzic—the architect of the campaign of ethnic cleansing—had been apprehended and put on trial. Bosnia was essentially divided into two hostile states—the Muslim-dominated Federation and the Serb-dominated Republika Srbska.

Though under the banner of one nation, the divisions were still strong, and politicians still tried to fan the flames of nationalism that burned just below the surface. Each of the two states had its own parliament, police force, and school curricula, its own laws and regulations. They even had their own separate armies. The central government had limited responsibilities and limited powers. NATO forces still handled much policing in the country, maintaining an uneasy peace between the two sides.

Provisions in the agreement that ended the war and created the two separate states within Bosnia attempted to set up mechanisms that would protect "human rights." A Human Rights Commission was established, which could make observations and issue reports, but had no power of enforcement. Refugee returns, arguably one of the most important factors in rebuilding society, depended on the ability of the refugees to trust the institutions of the state in their home area, especially the police force. Muslims who had been expelled were reluctant to return to Serb-controlled areas, and Serbs who had fled were reluctant to return to Muslim-controlled areas. Perhaps most troubling for children was the loss of friends from different ethnic groups who had left their homes and would not return.

The Dayton Accords, which ended the war in December 1995, set up a system that stopped the fighting but left the wounds created by the war largely unaddressed. There was no *ecological* approach to rebuilding the society—no effort made to address the underlying causes of the war, the interconnected effects of the war on a community level, or the circumstances that could lead to another one. Police forces still harbored former military and paramilitary soldiers; schools still taught their own version of history in which the victimized group was always the group writing the curriculum, in which there was always an enemy "other." The separate media rules and education systems ensured that the propaganda on each side would continue to flow. Refugees did not return, by and large, to unfriendly areas. The ethnic division of the country had been institutionalized.

Another war was not out of the question, but war was not, however, the major concern for the children on Mount Igman, neither memories of the war if they were old enough to remember, nor fears of a new war coming. They had more pressing concerns than politics and warfare.

"There is too much crime in the city," Elvira said. "There are too many drugs and no jobs. It is very hard in Bosnia, you know?"

Elvira was a high school student, a plucky girl who spoke English well, fretted about boys, and enjoyed looking after the younger kids on the mountain. More than once, I saw her scold Christof for his attitude toward the weaker or littler children. She was a bright, motivated youth, who, under different circumstances, would have found life brimming with opportunities.

"The problem is," she explained, "there are no jobs. I do not know what I'll do after school. Even with a university degree, there is no work." Bosnia has a 40 percent unemployment rate and young people are often anxious to leave. Elvira spoke German and hinted at the possibility she might go there, though Sarajevo was her home and she was not eager to leave it. Her family had stayed through the war, hiding in basements with their neighbors, dodging sniper fire to get water from the well or firewood to heat their home. Why would she leave now? She was conflicted and hoped the situation would change.

After coming down from the mountain and spending some time in the city, I sat on a rooftop with a group of men in their early twenties. We drank beer and looked at the burnt out towers of the twin skyscrapers, nicknamed Momo and Uzeir. Momo, a Serbian name, and Uzier, a Bosnian name, gave the towers significance before the war. They represented Bosnian unity. Because no one knew which building carried which name, both towers were destroyed in the siege. Now the corpses of the twin buildings stand as far more grim reminder of what brother can do to brother.

Next to us, the old police barracks still stood, though mortars had ripped it open and torn its guts out. The rooftop on which we sat took a hit and bore a huge gash in its side, the metal railing twisted and bent.

"They used to lob mortars at the station and hit our apartment building," Omar told me. "I lived on the other side of the building, so it was okay. Though we still went to the basement when the shelling was close. It was an exciting time."

Omar laughed and joked about the war. He was a young teenager during the siege and the whole thing had seemed like an adventure. He reminded me of Anna Freud's conjecture about the children during the bombing of London in World War II; that the young could not only handle but even enjoy the chaos of war, so long as it only threatened their lives and disrupted their routine, so long as the family stayed together. He and his family survived. For him, the war was a distant memory, one that did not trouble him greatly.

"I'm not traumatized," he laughed. "No more than anyone else in this crazy city." He took a drag on a cigarette and ran his hand through his dreadlocks. He wore a T-shirt that read *Don't Panic, I'm Islamic* that he got a great kick out of quoting every time we mentioned that I was American.

"The economy, man, that's my problem," he finally said, after giving some thought to the host of problems facing a young Muslim man in the Balkans. "That's what gives me nightmares. There's nothing to do in this town and no way to make money." He made a little, he explained, selling small bits of marijuana to his friends.

"There are some hardcore dealers in the city too, dangerous guys. They weren't always that way though. I went to school with these guys. But now there's no way out and no other way to make money, so they sell drugs."

Before the war, the suicide rate in Bosnia was around eleven per 100,000 people. Since the war, the rate has almost doubled, to around twenty suicides per 100,000 people in 2003. Discussing the problems of depression in Bosnia—psychological and economic—a woman who had lost her husband to a mortar told me with a shake of her head, "You know . . . we've survived the war. Now we have to survive the peace."

Her youngest son, a fifteen-year-old who had been a baby at home when the mortar came through the wall and killed his father, sat with us a while but felt no desire to talk about the past. He was more concerned with the present—the struggling economy, the war criminals still at large. He did not stay in the room long to talk, and he seethed with a bit of anger toward his mother when she spoke lightly of any times during the siege. He was not entirely comfortable with an American, his mother explained when he left to go out with friends. He suspected America was waging a war against all Muslims and thought I might be judging him, hating him secretly because he was Muslim. He may, however, just have been a shy teenager.

"After the war," she told me, "he always tried to protect me, keeping the shades drawn, avoiding windows. He could not understand why people fixed the glass in their windows or greenhouses . . . he thought they would simply be shattered again by mortars or grenades."

The aftermath of the war affected her youngest son deeply, and he remained terribly protective of his mother, even as she and I spoke. Before leaving, he ran me through an intense interrogation about who I was and why I was visiting. Though he spoke decent English and very good German, he refused to speak them in front of me, other than to ask what I was doing in Sarajevo, preferring to talk to his mother privately in Bosnian when I was present, even to have her translate, though he understood everything I was saying. His mother suspects that his anger and his trouble in school, despite his intelligence, go back to the loss of his father.

"He is very much like his father," she said. "I loved his father very much, and my son is stubborn just like him. He reminds me more and more of him every day." Her eyes grew moist, though she did not cry. She entertained countless journalists in her home during the siege, she told me, and had told her story many times. Telling her son's story, however, was harder for her. He, like teenagers the world over, was struggling to figure out what kind of adult he would be.

As poverty grows in Bosnia, as the post-war economy stagnates, young people are left with few options. Frustration grows daily and politicians on all sides seem incapable of making things work. The two parliaments constantly block each other's initiatives. Al Qaeda has begun to get a foothold in the region, forming safe houses in Sarajevo; right-wing clerics are gaining prominence. Several suspects have been arrested in Bosnia under terrorism laws, charged with planning suicide bomb attacks on American and European interests in the Balkans.

At the same time, the Serbian nationalists continue to honor and protect former war criminals. At the funeral for the mother of the former Serb leader, Radovan Karadzic, people carried signs and political banners proclaiming him and General Mladic heroes. This environment of high tension and little opportunity is a hard one in which to grow up, in which to heal. Everyone is tired of fighting, but peace has its own challenges.

Children manage the minefields of peace and of recovery from the trauma of war in radically different ways. Jaca engaged directly with her memories, telling stories, reading about politics, asking questions. Omar never thought much about it at all. Christof rarely spoke of the war, though the consequences of it loomed large in his interactions with the other children and with our dog.

I spoke with dozens of children in my time in all of the conflict areas I visited who did not want to talk about their experiences of war. Justin, the tall Rwandan boy who lost his mother and father, felt better because, as he said, "I am learning to forget."

I was shocked to hear that sentiment echoed by adults all around him and around many other children in similar situations. The adults were trying to forget as well. They stated proudly how their programs were designed to help the children forget the terrible things that had happened to them, to not think about them at all. Words like repression and avoidance buzzed around in my head. I came from a culture where forgetting was seen as pathological. People in America spend small fortunes trying to remember traumatic events as a way to become healthy again. My gut reaction was that this kind of avoidance would only lead to problems, but I'm not a trained psychiatrist. I said nothing. How else to cure trauma but by talking about it, I wondered. Avoiding the subject, repressing the memories of the terrible events of war, would lead these children to problems, I feared. Post-traumatic stress disorder. Pathological behavior. How could they move past the things they'd experienced unless they talked about them, exorcised the demons of their past, so to speak?

A thirteen-year-old Albanian boy named Peter had lost his father, who was shot by paramilitaries. He wore soccer shorts and a red tank top that needed washing. His face was downcast, and he did not join in the conversation with the other children on the day we met. He did not want to talk about the past or the future.

"They took my father and burnt him in a house with some other men, what does it matter?" he said, angry that I would pry, a total stranger. The principal of his school told me he was quite a good student, "well adjusted," though a bit sullen at times, like any thirteen-year-old boy. His family was poor.

"It makes sense he would be sad," the principal said. "If the ocean was a piece of paper and the sky was a pencil, that would not have been enough to write the suffering of the Albanians." The principal was also a published poet and had spent years in a Serb-run prison for his political verses.

Despite my fears at Peter's avoidance of past troubles, he was doing well under the circumstances. For him, avoiding the subject worked. Talking about it distressed him, being forced to remember and being helpless to change the past. He was not in denial, he just didn't want to talk about it.

"They took my father and burnt him in a house with some other men, what does it matter?" His words echo in my head, the matter-of-factness, the tight grip on reality.

"We don't talk about these things in my family," eleven-year-old Adem said. "Though I know that three houses in our village were burned down by the paramilitaries." He looked around the classroom where we sat and stared at Peter a moment. I asked the group of kids how they dealt with memories of the past, memories that were unpleasant. Adem answered for everyone. "We play to get over it." Peter nodded in agreement.

The *talking cure*, as it is sometimes called, is an individualistic concept. It fits well in the West, where the individual reigns supreme. But in other cultures, the individual is of far less importance on his or her own than the group, the community. In a society emerging from war, it is not just the individual who has been traumatized, it is society. Everyone has felt the sting, not just as individuals but as a community. The very institutions that make a community have been traumatized—laws, moral norms, family structure, economic security. Especially in cases of ethnic conflict, it is the group trauma that trumps all others.

In Bosnia and Kosovo, everyone talked about *what they did to us, what happened to the Serbs, what happened to the Muslims*, and so forth. Individual suffering was always set in the larger context of the suffering of a people. The children often saw their own suffering not as an individual event but as part of the campaign against everyone they knew or identified with.

In that light, healing for many of these young people could not happen in a vacuum. Their troubles were tied to the community's troubles. Their worries were not symptoms of psychological problems, but real reactions to the situations in which they lived. As Sigmund Freud wrote, "In an individual neurosis, we take as our starting point the contrast that distinguishes the patient from his environment, which is assumed to be 'normal.' For a group all of whose members are affected by one and the same disorder no such background could exist."

Safety and security tend to be the main concerns for young

people immediately after the end of a war, but as security threats lessened, access to education, better healthcare, and the means to make a living took the forefront. Social justice was also a concern that stayed on their minds. A sense of group victimization kept many of the children trapped in their wartime memories. Perhaps, for the young, it was better not to think about it.

Looking at the Albanian children of Kosovo, I saw that their sense of well-being depended on independence, the hope that history would not repeat itself. Their greatest fear was being tied to Serbia, once more at the mercy of the Serbs. For the Serb children, much of their anxiety came from their imprisonment in protected enclaves, their feeling that they were surrounded by a hostile nation, cut off from their own people in Serbia, left at the mercy of the Albanians. In Bosnia, crime and economic hardship dominated the worries of young people. They also worried about war criminals who were still at large, who were keeping populations in terror and keeping people from returning to the homes they lost in the war.

The problems of a post-war society are vast and complex and all of them are connected, especially in a society where ethnic conflict has torn everyone apart. As Kosovo looks toward independence and Serbia looks to block that hope, and as Bosnia tries to enter the European Union and bring its war criminals to justice, the region may once again find itself in turmoil, and many of the children who survived the wars of the 1990s will be the young adults who fight in the new conflicts. Meeting many of these young people, I cannot say if they will choose the path their parents took. I am sure many will. I like to think, however, of Mount Igman and the soccer games I played.

Every evening before dinner most of the kids would gather on that field littered with spent shell casings buried just beneath the dirt. The dog, Prijatelj, would trot along behind us, ignoring the occasional unkind word or tossed stone, and flop himself down under the shade of a tree to watch and to pant. The children, a jumble of different histories and dreams, would pick teams with the toss of a rock, or a coin if one was handy. They put aside their parents' violent history, their own troubled past, and they kicked the ball around, laughing and passing and dodging and slipping, the way kids do all over, in peace and war and poverty and riches. They played soccer together. They played every evening for as long as they could, at least until it grew too dark to play.

SEVEN

"God Has Something in Store"

What Becomes of War's Children

In the Autumn of 2004, I received an unexpected phone call from a young man named Joseph. He lived in Michigan, he explained, though he was not from Michigan. He came from a small town in southern Sudan near the White Nile. As a young boy, he had fled because of the bombings and attacks by the government army in Khartoum. He was separated from his family and wandered in the desert with thousands of other youths like himself, the youths who became known as the Lost Boys of Sudan. He had survived, he explained, through the grace of God.

He lived with the other boys, his brothers in suffering, he called them, in Kakuma Refugee camp in Kenya until, thanks to the U.S. State Department and American church groups, he was put on a plane and given a new life in an American city, resettled to a third country of refuge far from his homeland. He received basic training in what life would be like in America.

Simple things: *This is a flush toilet, this is a gas stove, this is a bread knife, a utility bill, a paycheck.* He was loaded onto a plane with hundreds of other boys, one of countless flights that left the refugee camp over the years during which the resettlement program was operational. It was his first time on an airplane. Until then airplanes had only dropped things on him, bombs or pallets of food aid (which could also kill someone crushed underneath). This acceleration away from the earth was something new to him, exciting and terrifying, a microcosm of the emotion he felt about abandoning his homeland and his people to start his new life in the United States. At twenty-three years old, America became the fourth country in which he had lived but the first that he did not arrive in on foot.

He had, however, left someone behind in Kakuma camp, he told me over the phone: his cousin, a young girl, his only family. He worried for her, all alone. He knew to contact me through his cousin; he had gotten my information from her, the young woman who chose to be called Rebecca.

"You have met her in Nairobi?" he asked me.

"Yes," I told him. "I met her one year ago." I didn't want to give him too much information just yet. I didn't really know who he was, calling me out of the blue. Rebecca had been in hiding, after all. A man was after her, a man who could have friends all over America through the resettlement program. Not all the Lost Boys were little angels, and most were grown men by now.

I knew that Rebecca was currently living in a protection area created by UNHCR, and her case was being reviewed for possible resettlement to the United States. Officials in Kenya with UNHCR and the State Department could not legally give me details about her case, but they told me she was anxious, not sleeping well, afraid of the others in the protection area. Her case, last I had checked, months earlier, was languishing in red tape, red tape that had gotten far thicker after 9/11. The United States was, reasonably, being very careful about whom it let in. They required countless immigration and security interviews, even after the UNHCR vetting process. Rebecca had told me when we met that she dreamed of going to live with her cousin, how she would continue to pray to be reunited with him. My heart leapt at the thought that this could be her cousin on the phone with me.

"I have news for you," he said and did not wait for me to ask what it was. "She has arrived in the United States. She has been resettled here, in this town, with a foster family. She will begin the tenth grade in September."

I don't remember clearly what I said or what happened next. I wasn't in my apartment anymore; I wasn't on the phone. I was back in Africa, under a blazing hot sun, walking to the taxi with Rebecca after our interview. The driver did not want to give her a ride to Kibera, the giant slum in Nairobi in which she was staying at the time, awaiting word from UNHCR if she would receive protection. Kibera is a slum of six hundred thousand people, more its own ramshackle city than a part of Nairobi. I did not want her walking or taking the bus. There was a man after her, a rapist and kidnapper. She glanced over her shoulder constantly.

"Is not safe to go to Kibera," the taxi driver said. "I do not drive there. They will rob you for sure." From what I could tell, he was a decent man, fair and intelligent, making an honest living in Nairobi. He wore a simple button-down shirt and blue slacks. His hairline was receding and flecked with bits of gray.

His concern for my well-being was generous though not really his job, I told him.

What he said next shocked me. He looked Rebecca up and

down, a withering stare, and then turned to me. He spoke quite loudly.

"Foreigners and terrorists live in Kibera. They come from Sudan and don't want to work. They just sit around and get drunk and chew qat." He shook his head and would not look again at Rebecca. He had no use for refugees, didn't want them in his city, didn't want one in his cab. They were backward, rural people exploiting the kindness of the Kenyan state, and, he implied, drug addicts to boot. He had no use for them. "Besides," he added. "They have no good roads there. It will break my car."

"I can look for another taxi if it's a problem for you," I told him. I had hired him for the day and not yet paid. He stood to lose some serious money.

"Fine," he grumbled. "I will take her to the edge. She may walk from there. I will not go in. It is not safe."

We drove through the city, past the university and the shopping areas, past the offices of international NGOs, past tall shining buildings and ramshackle lean-tos housing shoe repair stands, Internet cafes, odd assortments of clothing and cell phones for sale, and we arrived at the edge of Kibera. Rebecca got out of the car.

I wished her luck and gave her my card with my number written on the back and my personal e-mail address, told her to write me if she needed anything. She nodded, though in the space between us it was clear that there was nothing she needed that I could provide. The U.S. State Department, the Department of Homeland Security, and the United Nations held her fate in their hands now. She thanked me anyway and turned to walk off toward the dirt and mud streets of the Kibera district.

She looked back to wave as we drove off, and I knew that was the last I would ever see of her. The road I imagined her heading down was not a pretty one, but it was well trod by girls just like her, centuries of girls just like her, lost girls in vast slums all over the world. Too many girls whose lives ended at sixteen, too many girls who had no choices left, no doors open, no other paths. Patience and Charity and Hope with those crazy pseudonyms I gave them to try and tell their stories without ever really knowing their lives. Or Melanie or Nora or Thinzanoo, all these girls penned in by history and culture, by violent men, by entire governments with policies intended to erase them. I did not like the path I saw ahead for them or for Rebecca, who chose her pseudonym herself because it was a Christian name. I looked back for as long as I could, as if my seeing her would protect her, but I lost her in the crowd and we drove back to the hotel.

However, Rebecca's story did not end there.

"You hear this?" the voice on the other end of the phone exclaimed. Joseph. Rebecca's cousin. Reunited. I remember smiling, a smile that nearly knocked me over. "She will start school only one year behind her age. She is working very hard to speak English better now."

It would be almost two years before I spoke to Rebecca again.

"I am waiting," she said over the phone with street noises behind her, her English confident. "I am waiting for my registration paperwork so I will attend college." She lived in a small town in the Mid-Atlantic and was excited for all the opportunities ahead of her. She lived with her boyfriend, who would help support her as she went through school, though mostly it would be as it had always been for Rebecca: she would get by on her own.

Rebecca arrived in the United States in 2004, barely speaking a word of English. She was terrified and could not imagine going to high school, but she knew she had to do it to make something of her life, of the chance that had been given to her.

"I lived with a foster family and completed high school. It was strange, coming to America. I have a new life now. I experienced so many hard things, it was easy to give up, but God has something in store for all of us, I think."

She did not yet know what she would study, but she was thrilled to have options. She had made a transition, across the seas. She carried the lessons she learned through "disaster, pain, and difficulty"; she carried her faith and courage, and some hard-won street smarts. These were the sum total of the wealth she brought with her to make a life in America. I picture her on a college campus. The leaves are falling from the trees in browns and yellows, the same shades as the desert floor. She tosses her textbooks into a backpack, her hair tied up behind her head, a coat thrown on to shield against the foreign cold. She rushes to class past the kids on the lawn playing Frisbee, playing soccer. She has places to go, no time to play right now, but she can't help but slow her pace and look back over her shoulder a moment to watch the games unfold on the lawn. She's transformed from a child of war into a woman at peace.

Despite Rebecca's success in America, resettlement in a third country is not the solution for very many displaced children. A complex array of factors contributed to the massive resettlement program of the Sudanese youth, including pressure from religious groups, media attention and U.S. strategic interests in southern Sudan.

These factors do not exist for most displaced populations, and that aside, ethical considerations make third country resettlement problematic. Many southern Sudanese wondered aloud to me how their society would function with such vast numbers of their youth overseas. The hope is that the educated "lost children" will return and bring their skills and education back to Sudan with them when the violence ends. Many send money back to the refugee camps to support friends and family. Still others feel cut off from their people and find it difficult to adjust to a new and alien culture, suffering bouts of restlessness and depression.

Separating children from their culture and maybe even from family that may still be living is no one's ideal way of protecting youth during war. Aid programs aim by and large to keep families together and to strengthen a family's ability to care for itself. Often it is mothers who are the nexus of aid and assistance for their children and, when dealing with orphaned or separated children, much effort is made to find them foster parents within their communities. Rarely do affected populations think resettling children to third countries is a good solution, though sometimes, as in the case of the Lost Girls, more danger comes from their families and communities than from the war itself. That aside, however, the ideal for children of war is not exile but homecoming.

When I visited Kosovo and Bosnia, which I knew would be my final trip for this project, I brought along a book I should have read in college, Virgil's *Aeneid*. The hero, dutiful Aeneas, escapes the burning city of Troy as it falls and sets off into exile, destined to reach a foreign shore that his progeny will call Rome.

On the night of the destruction of Troy, Aeneas races through the city, dodging marauding Greeks, crashing rubble, and flames—his account sounds oddly similar to the children's accounts of fleeing their villages and towns. With tears in his eyes, he seeks his wife, his father, and his son, hoping to rescue them from the terror that will come at the hands of the city's captors. He finds his father and his son, but his wife's spirit appears and urges him to flee, tells him that she is already dead. He obeys and arrives at the shore to behold a woeful sight, one that I imagine every refugee entering a camp or escaping a conflict has seen at some point.

Aeneas speaks: *Here I find, to my surprise, new comrades come together . . . joined for exile, a crowd of sorrow. Come from every side, with courage and with riches, they are ready for any lands across the seas . . .*

When I look at this passage, I cannot help but think of the multitude of war's children I have met, the littlest exiles, a tiny citizenry shuffling below the radar of history, and I think of the journeys on which they have embarked, the riches they carry with them, and the shores they strive to reach.

I think of Paul, exiled not just from his village, but from his childhood, pushed headfirst into adulthood as a soldier, longing to return to school. Paul who loves peace could, by now, be a soldier again. At eighteen years old, no one could stop him. I asked around in the region. No one knew what became of him, no one knew if he had the chance to go back to school. I like to picture him a mechanic in a peaceful city, grease on his hands and that same glowing smile he graced me with. But I've been stopped at checkpoints, and I can imagine that same smile terrorizing those at its mercy. Paul understood survival, and if survival meant a return to violence, Paul could have returned to it. How many times in his young life had he seen the power dynamic play out with brutality? How many times had he seen the grace of mercy? The math isn't simple. But I remember when he told me that he could not be afraid, and I hope his courage gave him the strength to grow toward peace.

Keto's an adult now too. Thousands of children in Lugufu camp took their secondary school graduation tests and thousands returned to the Congo ahead of the first free elections. Perhaps Keto was among them, a full-fledged citizen, a minor no longer, able not only to choose his own destiny but the destiny of his country, something previous generations had never fathomed. Keto was smart and could, more than likely, play the system well, perhaps become a Big Man, perhaps a doctor or a teacher. He could also be wallowing in that camp still, pining away the days in search of rescue, unable to wean himself from the habit of dependency on aid and unable to find a job in a nonexistent economy. No way to tell. My translator and I arranged to get him his treasured soccer ball after I left, and I heard later he was leading games on the dirt field near the school. That was the last I heard of him, calling out passes and field positions, guiding his team, even if there was no keeping tally of the points.

Christof did not return to the summer camp on Mount Igman; he stopped attending the Sunday program. In some neighborhood in Sarajevo he continued to work out his anger and his gentleness, figuring out which side of himself to follow. Bosnia seems poised for a hard time ahead, and young men with little money and a lot of rage could be the powder keg that sets off another war. As the most senior international official in Bosnia noted, the situation "can all too easily escalate into violence in a society where weapons are everywhere, alcohol plentiful, and the summer long and hot." One likes to think that when the moment comes, Christof will remember the dog, Prijatelj, panting under the heavy sun on Mount Igman, that he will remember giving little Sofya a ride on his back, and how peacefully she slept with her arms around his neck. One hopes that he will join with other moderates and be a voice for cooperation. These decisions come from the fabric of a life, after all, and there is much sewn into the fabric of his life, of all their lives.

The riches each child of war carries into adulthood are forged from moment to moment; who they will be is built up quietly as events unfold, as they try to fathom the kindness they receive and the betrayals they feel, as they calculate their survival and learn who their friends are.

The young experience it all in war: the highs of human kindness and the lows of human cruelty. They are at both ends, giving and receiving, their stories unfolding without fanfare. Records are rarely kept on their movements and little notice is taken of their deaths unless it serves a political end. There is no prescription, no single way to ensure that children survive the hardships of war or flourish as adults but to eliminate wars altogether. Granted, for some children, the challenges they face in times of crisis can actually benefit them, help them build "character," as I saw time and again among the children I met. But there is no way to know that they would not have flourished in peace-time as well, tackling the day to day challenges of living with the same energy and courage as they tackled survival.

If a society restores itself as part of the peace process, rebuilding its institutions and moral norms, if it includes the young in the dialogue of rebuilding and renewal, the odds for war's children improve. Sometimes greater intervention will be needed— psychiatric care, physical therapy, job training, development aid, peacekeeping forces. Each culture and each conflict are unique, as is each child. They want different things for the future.

Jeanine, a fifteen-year-old Burundian girl sleeping on the streets of a refugee camp, wanted peace for her own country and the opportunity to move home again, to rebuild. "I want to return home, when it is safe," Jeanine said. "I want to go home and farm my land."

She scoffed at the idea of moving to the United States or Canada as many others from the camp longed to do. She wanted to go home. She did not want to live the life of an exile nor the life of handouts that refugees are subject to, and she had had enough of parents. Her own were dead and the ones who took her in, her foster parents, she said, mistreated her. The war in Burundi, the war that took her parents from her, did not define her, though it impeded her dreams. It was a distant memory and the day-to-day things, the chores, the gossip, the games, the work of living took up her time and her energy.

Paul too scoffed at the idea of moving to the United States or Canada, though he did indicate that he would like to be somewhere safe where he could attend school.

"Can a child do this in Canada?" he asked. Perhaps his reluctance to resettle if he had the option was due more to a lack of perspective than a lack of desire. Regardless, his main desire for the moment was to get an education and to get out of the demobilization center where he lived with the other boys. He also said he would like a real soccer ball. The other boys at the center seconded his wish. Their dreams for the future ran parallel to their quotidian desires: soccer and joking and scribbling in books and drawing pictures and still more soccer.

Marko, the ringleader of his group of friends in the Serb enclave in Kosovo, wanted to return to Pristina, the capital city, where he had lived before the war. He was tired of the provincial life and longed for the hustle and excitement of the city, which was cut off from him by ethnic conflict. He did not want to leave Kosovo, as his parents often mentioned.

"Kosovo is our place," he said. "We don't want to leave. We want to remain here and to remain part of Serbia." Asked if he wanted that even if it meant another war, he toned down some of his bravado. "No one wants war," he said. Having seen his displays of kung fu against hypothetical Albanians, I wondered, when he got older, which version of Marko would come to the foreground, the one who wanted peace or the one who was ready to "kick Albanian asses"? The question is hardly academic. War's children will one day become the adults of their societies.

There is no way to know for certain what sort of adults they will become. Their actions probably suggest more about the moment in which they act; their inconsistencies the working arithmetic of building a life and of surviving. They showed me parts of themselves, the parts they wanted to show. I saw other parts of some of them when they let their guard down during a game or a long walk, as when Christof slipped our dog some water and patted him behind the ears when he thought no one was watching. Other parts of who they are I've guessed at, based on their drawings, on what other children and other adults told me, based on what I've learned about their history.

Christof's cruelty and his kindness do not exclude each other, nor do they sum him up. Marko's wish for peace and his inherent racism, though contradictory, hold the key to what he will one day be, though no one, least of all Marko, knows what that is yet. Paul's capacity for violence and his generosity of spirit will vie with each other in his identity, his memories and his values often at odds.

In *Civilization and Its Discontents*, a book written when the possibility of a second war in Europe loomed large overhead, Sigmund Freud described the two primal drives in human beings: *Eros*, the instinct that drives us toward love, to seek out comfort and support (and pleasure) from others, and *Thanatos*, the instinct that drives of us toward death and destruction, the instinct that pushes us not to love our neighbor as ourselves, but to use our neighbor, exploit him, rape him, humiliate him, and burn his house to the ground. He implies that these two instincts are the two "Heavenly Powers" battling for supremacy: good and evil, God and the Devil, Life and Death. He ends his brief essay on human civilization with a passage added after publication to the second edition, as his own anxieties about Hitler's rise to power grew stronger.

"Men have gained control over the forces of nature to such an extent that . . . they would have no difficulty in exterminating each other to the last man," he wrote. "And now it is to be expected that the other of the two Heavenly Powers, eternal *Eros*, will make an effort to assert himself in the struggle with his equally immortal adversary [Death]. But who can foresee with what success and with what result?"

Looking for clues on what the future might hold for war's children, I am reminded of an incident that occurred early in my travel. I had wandered to the school in the part of Lugufu camp where Justin, Keto, and Melanie lived. The school consisted of three low-slung buildings with thatched roofs and dirt floors. The buildings surrounded a large dirt field, and the children spilled out onto it the way children always spill out of classrooms when they are released: with a lot of shouting, shoving, and giggling. I kept my distance, not wanting to change anyone's behavior by my presence. I watched as girls and boys ran off into the rows of houses and tents in the camp, holding hands and laughing. I saw little Melanie, wearing the same tattered red dress she wore when we met. She chased after two larger girls, all of them laughing, and disappeared into the camp.

A group stayed behind to play soccer, this time with a real ball that belonged to one of the teachers. The moment he took it out of his bag, the boys crowded around him. He shouted out commands, dividing them into teams. This game was more organized than most and would, it seemed, have goals and points. The teacher acted as referee. He set the ball down, and the game began. The teacher strolled from side to side, calling out what sounded like reprimands or advice. He never stopped the play, even when one boy tripped Keto and Keto got up to shove him. They wrestled a moment, with everyone around watching, and the scuffle ended nearly as quickly as it started. Both boys returned to playing. They seemed to be on the same team.

I saw Justin standing on the sidelines, not playing. He saw me and wandered over to where I stood. Several boys followed. The teacher noticed me, but exhorted the boys to keep playing.

"You play football?" Justin asked, pointing at the field.

"Not today, I think."

"Me not today too," he said. His English was not half bad. The other boys around us whispered and poked each other, staring at me, wondering what Justin and I were saying.

"They never talk to a *mzungu* before," he explained, smiling. We had spoken the day before and he was, therefore, an expert. "You first time to Africa?"

"Yes it is."

"This is our school. We have—"

"Greetings!" the teacher trotted over, leaving the game to its own devices. The children kept playing as the referee quit the field and came up to me, interrupting Justin as if he weren't there. "You are welcome."

"Thank you."

"This is our school," he explained and offered to give me a tour. I accepted, and we walked together, Justin accompanying us and the other boys walking just behind.

"Your first time to Africa?" the teacher asked.

"Yes."

"You are with an NGO? Or UNHCR?"

"I'm researching the lives of children."

"The children have it very hard in this camp. I try to teach them, but there is no money. I do not get paid, you know?"

"I did not know that."

"And we have no money for materials."

"Where do you get your funding?"

"The school was built by the UNHCR, but we get little bits of money here and there from the community. It is not very much."

"Justin was just telling me about the school."

"Justin, yes," the teacher said, acknowledging the boy for the first time. He patted him on the shoulder. He said something to the other boys, and they answered. Justin looked at his feet. The teacher and Justin exchanged words briefly, while the other boys watched. I did not speak the language, but I could tell when a teacher lectured a pupil. He had the downcast look, part misery, part defiance. I looked to my translator.

"He tells him that it is time for adults to speak and it is not right to listen," the translator whispered. "He tells him to go play like the other boys." With a quick pat, he sends Justin running off toward the game with the rest of the children.

"This boy is troubled. I try to make him forget," the teacher said. "He does not play like the others and he thinks about the past very much."

"Maybe this is his way of coping," I suggested.

"No, I know this boy," the teacher said, watching him join the game. "He will not be well if he does not play with the others."

We watched the game together for a while, not saying much. Justin played reluctantly, hanging back, never rushing in to seize the ball. The others largely ignored him. As he played, he looked over at where we stood, his longing to return to our conversation obvious. He did not fit in, though his teacher was determined that he try to fit in.

I suggested Justin might not want to play soccer anymore.

"The children must be part of the community." He said no more about it. Sigmund Freud also wrote in *Civilization and Its*

Discontents: "There is indeed another and a better path [to happiness]: that of becoming a member of the human community." I cannot imagine this teacher had read Freud, but through his own observation he came to the same conclusion as the founder of psychoanalysis. In order to survive, one must be part of something. We are social creatures; we cannot do it alone. Not he, not I, not little Justin.

Within a few minutes, Justin had the ball and maneuvered it down field. One of the other boys knocked it from him. He chased after it and kicked at the boy's legs. The boy stopped playing, threw a punch that connected with Justin's shoulder. Justin smacked him back. They scuffled a moment, and the teacher beside me simply watched. The second fight of the game was brief and dusty and when it ended, the game resumed. Justin, visibly angry, kept playing. Within minutes he was part of the game again, smiling and shouting with the rest of them, having what looked like a good time, the scuffle forgotten, though his eyes flashed with anger whenever he passed near the boy who punched him. But the others seemed, for the moment, not to care that he was an outsider, a Tutsi among the Hutu, an orphan among the parented. They played together because that was the task at hand. Though Freud's "inclination toward aggression" emerged for a moment, it was a form of love, the *Eros* of play, that won the day. The children played and kept the drive toward death at bay, at least for a few more hours. Gunfire often ripped the night air in the refugee camp, *Thanatos* giving the children no quarter, even as they slept.

War's children are not a lumpen mass, but a collection of little hopes and needs and impulses and desires lived out from day to day. They live their lives amid the backdrop of terrible violence and deprivation, amid constantly shifting loyalties and labels and dangers, but also amid the backdrop of going to school, of who-can-juggle-the-soccer-ball-better, of sewing torn pants, of funny pratfalls, and of family and friends. Of play. The children are a collection of all the things that happen to them, and the kinds of people they become is being decided every second, in the struggle of love versus death, in the battling of the primal urges and the relationships that form around them.

The most important factor for children in war is building a day-to-day life, having somewhere to go, something to do, someone to count on, seeing *Eros* in action rather than its destructive brother. Justin played soccer and, though it would not change a thing about the reality of the world in which he struggles, the minutes of play and happiness will weave themselves into his life, just as the horrors of death and destruction have woven themselves into his life. They are all part of his story, the narrative of himself he shows to others, and the secret history he shows to no one.

Children can survive without comforts—they are amazingly adaptable. They can survive without safety, even, drawing on what resources they have to get by, but they cannot long survive without hope. It is the job of every adult, those who make wars and those who watch them unfold, to bend their mind to the task of giving hope, of creating a world where childhood can flourish, where play is possible.

Embarking on this project I was amazed to find such a great crowd of remarkable children, struggling with the quotidian and the extraordinary, each in their own way. They have different levels of control over their lives, and they are all, to some extent, at the mercy of the adult world, subject to the policy shifts of governments and the mood swings of their caretakers, even their own contradictory impulses, as all children are. But they take control where they can, playing and dreaming and observing the goings-on about them. They remember what is done to them and what they have the power to do to others, and they will wield that power in the future when the world of the adults is their world, for good or for ill, to create or to destroy. The capacity for either is in all of them. *Eros* and *Thanatos*, Love and Destruction, fighting it out inside every one of war's children, the soldiers and the students and the soccer players.

After Justin, after Keto, after Christof and Nora and after Rebecca, after all of them are gone, whether they died still young or grew into adulthood, whether they were celebrated or feared, lost to the jungles, wandering from place to place, dispossessed, or brought inside to a warm bed or shot at or disemboweled or turned into killers or bandits or whores, whether they worked the land their parents worked or whiled away the evening hours kicking a soccer ball under an acacia tree humming an old song they heard when they were younger, whatever became of this one group of war's children, there will be new children and new wars.

The children of these wars, like those who came before them, those of whom they have no knowledge, will bear their riches inside themselves as well. Like Justin and Keto and Christof and Nora and Rebecca and all the others, they will face the dangers that they must, bearing their daydreams and ideas, their faith, their sense of play along with them. They will bear them into exile. They will bear them back again. Occasionally, they will count among their riches a jump rope or a pastiche soccer ball, or a tattered French grammar book.

They will also have burdens to carry. The burdens that grown-ups have placed on them: ethnicities and histories, violence and politics, hunger and poverty. The past, the present, and the future. The children carry what the adults put down. They will carry these riches and these burdens across bomb-scorched deserts, through deadly jungles, and down bullet-pocked streets. They will carry them into adulthood, if they survive. Some will not survive.

One boy haunts me more than any others. The night after Mount Nyiragongo erupted in Goma, I found myself in Kigali, Rwanda. I left the hotel where I was staying just across the street from the famous Milles Collines to exchange some money and to get a bite to eat. I walked up the hill toward the center of town. The road was packed dirt, and wide, deep gutters ran along the side. Palm trees gave a canopy of shade all the way to the town center, and it was a very pleasant walk as there were few cars. It was eerily quiet for a capital city. As one young woman described it, Kigali was a city with more ghosts than people.

To get to the Indian restaurant that also worked as a grocery store and informal marketplace, I cut across an empty lot. As I crossed the lot, I saw a boy of about twelve years old. I had seen him before, several weeks earlier when we first came through Kigali. He had begged at the Indian restaurant, and the owner had chased him away with a stick. He and I were alone in the vacant lot now, no owners, no sticks. He wore torn blue jeans that had been cut into shorts and flip-flops to protect his feet from broken glass and sharp rocks. His shirt was a tattered tennis shirt, light blue, torn open all the way to the bellybutton and filthy. He wore a white hard-hat that made him look almost clown-like, and his face was all smiles.

"*Mzungu*, mister. *Mzungu*, mister," he said and came toward me with his hand outstretched. He turned his lips down immediately in an expression meant to look pitiful. I was alone and had no way to communicate with the boy except for broken French and hand gestures. "Hungry," he said and repeated it again and again. "Hungry, hungry, hungry." I gave him some change I had, and he patted me on the back, transforming once more to smiles, tipping his hard-hat back on his head. "Hey friend," he said. "Friend man."

He laughed, a staccato laugh that shook his body. He seemed suddenly dangerous, though he was reed thin and several inches shorter than me. He pulled a cardboard box from his pocket. Glue. Cheap glue. The preferred drug of street children in this part of the world. He took a deep whiff, and his eyes went glassy, like big black marbles or tiny vacant cow's eyes. His face turned blank.

"My friend," he said again and nodded. Two men entered the lot at the far end. They looked tattered, but not nearly as frail as the boy. They stopped and took me in a moment, a young white man standing in a vacant lot talking to a drugged street kid.

"Hey!" one of them shouted, but I left the lot quickly, not wanting to find myself alone and outnumbered along what began to seem an unwise shortcut. I kicked myself for not doing more to help the boy, for not knowing what to do.

That night, walking with two colleagues back toward the hotel from the very same Indian restaurant, I saw the boy again. He no longer had his white helmet on, and I wondered where it had gone. He followed us down the hill slowly, calling after us, trying to catch up.

"*Wazungu!* Hungry! Hungry! Hungry!" His voice sounded pitiful. It was dark out and we were rushing to get back to the hotel, to get to sleep, to get on a plane the next morning. We did not stop.

I knew, as all of us did, that we were actually rushing away from the boy. After our time in the Congo and after the volcanic eruption just a day earlier we were exhausted, and not only physically. I have to admit to a deeper kind of exhaustion that night in Kigali. I could not bear to face another begging child, especially this glue-sniffing child with whom I'd shared one brief moment of connection. I couldn't bear to face my own helplessness that would be reflected back at me through his eyes.

"My friends," he called again, his voice reaching a new high pitch, practiced and pitiful. It hurt to hear it. "Hungry! Hungry!" And then, almost in a whisper: *"Help me. Hungry."*

He whined after us, his French vocabulary limited but confident. He followed like an injured dog, effacing himself of his humanity, whimpering. He did not care. Dignity would not get him the money he needed. Pity would. He knew the effect he was having on us. He was a skilled professional.

We walked faster, our hearts breaking. Then the dirt beside us kicked up in the air as if a bullet had struck. We ducked and looked back. I thought of the two other men from the afternoon, thought of a gang. For a moment, fear erased my guilt. But it was just the boy, no longer walking. He stood still on the road, his hands clutching a pile of small stones. The boy had thrown a rock at us. He raised his arm to throw another. It hit the ground near one of my companion's feet, kicking up another spray of dirt. The boy cocked his arm back to throw again, but we yelled at him to stop, turned toward him aggressively, and he ran off into the dark without a sound.

I thought of the owner of the Indian restaurant chasing the boy with a stick and how I judged him for his lack of compassion. I thought of Rudyard Kipling's *Kim*, a wily street kid in India, and of Charles Dickens' urchins in London. I did not think about that boy himself anymore that night, just the type of boy I imagined him to be. We got back to the hotel, slept soundly, and left Rwanda the next morning.

I do not know what happened to this boy. I never learned his name, and I doubt he lived long. I think of him now, the nameless street child, among so many nameless children, and I wish I had spoken to him, wish I had learned his story, heard his words in his own language, his memories, his day to day strategies, his ideas about his place in this world. I wish I had asked his name. I am sure he held a great deal of hurt within him, but also strength and courage, and also ruthlessness, and regret. I am sure he had done awful things to people and had worse things done to him. I am sure his tale was epic as any Aeneas, but he ran off into the dark, and I never had the chance to ask him.

REFERENCE LIST

BOOKS

Apfel, Roberta J.; Bennett Simon, eds. *Minefields in Their Hearts: The Mental Health of Children in War and Communal Violence*. New Haven: Yale University Press, 1996.

Ariès, Philippe. *Centuries of Childhood: A Social History of Family Life*. Trans. Robert Baldick. New York: Alfred A. Knopf, 1962.

Berkeley, Bill. *The Graves Are Not Yet Full: Race, Tribe, and Power in the Heart of Africa*. New York: Basic Books, 2001.

Coles, Robert. *Doing Documentary Work*. New York, Oxford: New York Public Library and Oxford University Press, 1997.

———. *The Moral Life of Children*. New York: The Atlantic Monthly Press, 1986.

———. *The Political Life of Children*. New York: The Atlantic Monthly Press, 1986.

Dallaire, Roméo. *Shake Hands with the Devil: The Failure of Humanity in Rwanda*. Toronto: Random House Canada, 2003.

Fink, Christina. *Living Silence: Burma Under Military Rule*. Zed Books, 2001.

Freud, Anna. *The Writings of Anna Freud: Infants Without Families*. Volume 3. International Universities Press Inc., 1973.

Freud, Sigmund. *Civilization and Its Discontents*. 1930. Trans. James Strachey. New York: W.W. Norton & Co., 1961.

Gourevitch, Philip. *We Wish to Inform You That Tomorrow We Will Be Killed with Our Families: Stories from Rwanda*. New York: Picador, 1999.

Heins, Marjorie. *Not in Front of the Children: Indecency, Censorship, and the Innocence of Youth*. New York: Hill & Wang, 2001.

Hochschild, Adam. *King Leopold's Ghost*. New York: Mariner Books, 1999.

Jones, Lynne. *Then They Started Shooting: Growing Up in Wartime Bosnia*. Cambridge: Harvard University Press, 2004.

Maass, Peter. *Love Thy Neighbor: A Story of War*. New York: Knopf, 1996.

Machel, Grace. *The Impact of War on Children*. New York: United Nations Publications, 2001.

Malcolm, Noel. *Bosnia: A Short History*. New York: New York University Press, 1996.

Mandelbaum, Allen. *The Aeneid of Virgil*. New York: Bantam Books, 1971.

Marten, James, ed. *Children and War: A Historical Anthology*. New York: New York University Press, 2002.

Marten, James. *The Children's Civil War*. Chapel Hill: University of North Carolina Press, 1998.

Nzongola-Ntalaja, Georges. *The Congo from Leopold to Kabila: A People's History*. London: Zed Books, 2002.

Postman, Neil. *The Disappearance of Childhood*. New York: Delacorte Press, 1982.

Rashid, Ahmed. *Taliban*. New Haven: Yale University Press, 2001.

Rosen, David. *Armies of the Young: Child Soldiers in War and Terrorism*. New Brunswick: Rutgers University Press, 2005.

Rosenblatt, Roger. *Children of War*. New York: Anchor Press/Doubleday & Co., 1983.

Scroggins, Deborah. *Emma's War*. New York: Vintage Books, 2002.

Singer, Peter W. *Children at War*. New York: Pantheon, 2005.

Sluka, Jeff. "The Anthropology of Conflict." In *The Paths to Domination, Resistance, and Terror*. Ed. Carolyn Nordstrom; Joann Martin. Berkeley and Los Angeles: University of California Press, 1992.

Ung, Loung. *First They Killed My Father: A Daughter of Cambodia Remembers*. New York: HarperCollins, 2000.

Werner, Emmy E. *Reluctant Witnesses: Children's Voices from the Civil War*. Boulder: Westview Press Inc., 1998.

Westermeyer, Joseph, and Karen Wahmanholm. "Refugee Children." In *Minefields in Their Hearts*. Eds. Roberta Apfel and Bennett Simon. New Haven: Yale University Press, 1996.

Williams, Elizabeth McKee. "Childhood, Memory, and the American Revolution." In *Children and War*. Ed. James Marten. New York: New York University Press, 2002.

NEWSPAPER/MAGAZINE ARTICLES

Amornviputpanich, Punnee. "Daddy Htoo: The Teen Twin Rebels Grow Up." *The Nation* (Bangkok), May 26, 2004.

Brennan, Richard; Husarska, Anna. "Inside Congo, an Unspeakable Toll." *Washington Post*, July 16, 2006.

Derluyn, Ilse, et al. "Post-Traumatic Stress in Former Ugandan Child Soldiers." *The Lancet* 363, May 15, 2004.

DRC: From Protection to Insurgency: History of the Mayi-Mayi. IRIN News Service, 2006.

Harnden, Toby. "Child Soldiers Square Up to U.S. Tanks." *London Telegraph*, August 23, 2004.

Peterson, Scott. "Child Soldiers for the Taliban? Unlikely." *Christian Science Monitor*, December 6, 1999.

Plemming, Sue. "Children Bear Brunt of Lebanon-Israeli War." Reuters, August 15, 2006.

Singer, Peter W. "Western Militaries Confront Child Soldiers Threat." *Jane's Intelligence Review*, January 1, 2005.

Subotic, Tanja. "Suicide Rate Doubles in Post-War Bosnia." AFP, May 6, 2003.

West Africa: Children in Danger; Begging for Teachers. IRIN News Service, 2006.

REPORTS/PAPERS

Apple, Betsy; Veronika Martin. *No Safe Place: Burma's Army and the Rape of Ethnic Women*. Refugees International, 2003.

Boyden, Jo. "Social Healing in War-Affected and Displaced Children." University of Oxford Refugee Studies Centre. AsylumSupport.info, 2003.

Caouette, Therese M., and Mary E. Pack. *Pushing Past the Definitions: Migration from Burma to Thailand*. Open Society Institute, 2002.

Colombia's War on Children. New York: The Watchlist on Children and Armed Conflict, 2004.

Democratic Republic of the Congo: Demobilization Programs Require Special Focus on Girls. Washington, D.C.: Refugees International, 2006.

Democratic Republic of the Congo: Reluctant Recruits: Children and Adults Forcibly Recruited for Military Service in North Kivu. New York: Human Rights Watch, 2000.

Edgerton, Anne. *Child Soldiers in the Eastern Congo.* Washington D.C: Refugees International, 2001.

Fitzgerald, Mary Anne. *Throwing the Stick Forward: The Impact of War on Southern Sudanese Women.* African Women for Peace Series: UNIFEM and UNICEF, 2002.

License to Rape: The Burmese Military Regime's Use of Sexual Violence in the Ongoing War in Shan State. Chiangmai, Thailand: The Shan Human Rights Foundation & The Shan Women's Action Network, 2002.

Making the Choice for a Better Life: Promoting the Protection and Capacity of Kosovo's Youth. New York: Women's Commission for Refugee Women and Children, 2001.

March 2000 Delegation to Sierra Leone Preliminary Findings and Recommendations. Physicians for Human Rights, 2000.

Myanmar: Atrocities in the Shan State. London: Amnesty International, 1998.

Pillsbury, Allison Anderson; Jane Lowicki. *Against All Odds: Surviving the War on Adolescents. Promoting the Protection and Capacity of Ugandan and Sudanese Adolescents in Northern Uganda.* New York: Women's Commission for Refugee Women and Children, 2001.

Protection Through Participation. The Women's Commission for Refugee Women and Children, 2003.

Shukla, Kavita. *Ending the Waiting Game: Strategies for Responding to Internally Displaced People in Burma.* Washington D.C.: Refugees International, 2006.

Total Denial Continues. Seattle: Earth Rights International, 2000.

Under Orders: War Crimes in Kosovo. New York: Human Rights Watch, 2001.

Unsettling Moves: The Wa Forced Resettlement Program in Eastern Shan State. Chiangmai, Thailand: Lahu National Development Organisation, 2002.

Untapped Potential: Adolescents Affected by Armed Conflict: A Review of Programmes and Policies. Women's Commission for Refugee Women and Children, 2000.

Youth Speak Out. New York: Women's Commission for Refugee Women and Children, 2005.

FURTHER RESOURCES AND WHAT YOU CAN DO TO HELP

It is easy to feel despair when looking at the problems faced by children in wars, but there are many programs that try to help and there are many resources available to learn more about these regions or issues, in addition to the materials listed as references.

Refugees International was founded in 1978 in response to the crisis in Cambodia and has grown into a powerful independent advocacy organization. Refugees International generates lifesaving humanitarian assistance and protection for displaced people around the world and works to end the conditions that create displacement. They accept no public money in order to remain an independent voice for action. Sixty percent of the proceeds of this book benefit their work. You can learn more about what they do at http://www.refugeesinternational.org. Their Web site also allows you to get involved through donations, hosting an event, contacting policy makers about issues concerning displaced and vulnerable people, and writing editorials for your local newspaper.

In addition to Refugees International, the Coalition to Stop the Use of Child Soldiers (http://www.child-soldiers.org/) works to prevent the recruitment and use of children as soldiers, to secure their demobilization, and to ensure their rehabilitation and reintegration into society.

Save the Children (http://www.savethechildren.org) helps children in crisis and children in poverty around the world. They run many community outreach, health, education, and recreation programs across Africa, Asia, and the Middle East. Save the Children also offers teacher's guides for discussing these subjects in the classroom.

CARE is a global poverty fighting organization, whose work focuses primarily on empowering women and on delivering aid to all people affected by armed conflict. You can become part of their fight by joining the CARE Action Network and making your voice heard (http://can.care.org/).

The Women's Commission for Refugee Women and Children (http://www.womenscommission.org) coordinates the Watchlist on Children and Armed Conflict, which advocates on behalf of the children of war on a policy level and publishes thorough reports on the state of children around the world affected by armed conflict.

Additionally, Human Rights Watch (http://www.hrw.org) investigates the conditions of children around the globe and advocates for their freedom, safety, development, and dignity.

By far the largest organization that helps refugees and displaced people is the United Nations High Commissioner for Refugees (http://www.unhcr.org). UNHCR is mandated to protect refugees, displaced persons, stateless persons, and all other persons under its mandate, to coordinate efforts all over the world in order to solve the problems faced by the displaced people and ensure respect for their fundamental rights: the right to employment and to education, the liberty of worship, the right to travel, and the protection of the law.

If you would like to learn more about refugee issues, there are many useful books and documentary films available, in addition to the material in this bibliography. These titles should be available from your local library. For a staggeringly thorough account of the contemporary phenomenon of child soldiers, Peter W. Singer's book, *Children at War*, is an invaluable resource. *A Bed for the Night: Humanitarianism in Crisis* by David Rieff examines the development of humanitarian organizations from their beginnings as organizations devoted to the alleviation of suffering to their current, more partisan state, and Michael Maren's *The Road to Hell: The Ravaging Effects of Foreign Aid and International Charity* looks at the effects of international aid on the societies it is meant to serve. *Darfur: A Short History of a Long War* by Julie Flint and Alex de Waal details the history of Darfur: its conflicts, and the designs on the region by the governments in Khartoum and Tripoli. It investigates the identity of the infamous "Janjaweed" militia and the nature of the insurrection, charts the unfolding crisis and the international response, and concludes by asking what the future holds in store. *Shake Hands with the Devil: The Failure of Humanity in Rwanda* by Lieutenant-General Roméo Dallaire with Major Brent Beardsley offers an eyewitness account of the failure to stop the 1994 genocide in Rwanda. A more detailed reading list is available from the Refugees International Web site.

As for films, Megan Mylan and Jon Shenk's documentary account of the Lost Boys coming to America, *Lost Boys of Sudan*, gives an insightful look at the resettlement process as it is experienced by Sudanese youths.

Burma: State of Fear, a film by FRONTLINE/World reporter Evan Williams, provides a look into the repressive nation of Myanmar. Williams traveled undercover to Burma (also known as Myanmar) to expose the violence and oppression carried out by Burma's government against its own people.

Total Denial by Milena Kanev, is the inspiring story of fifteen villagers from the jungles of Burma whose quest for justice eventually lead them to bring suit in a U.S. court against two oil giants—UNOCAL and TOTAL—for human-rights abuse. The filmmaker's "guide" during this journey was Ka Hsaw Wa, described by Kerry Kennedy in her book *Speak Truth to Power* as "a man of incredible courage and commitment, with the firm belief that one man can make a difference."

A sensitive look at the lives of three different children around the world can be found in *Living Rights* by Duco Tellegen, which explores dilemmas facing three young people on three different continents.

One of the most affecting films relating to issues of war and forgiveness is *Videoletters* (http:www.videoletters.net), which creates lines of communication between former enemies in ethnic conflicts. Beginning in the Balkans, the project has now spread to Rwanda and the volatile Caucasus region.

Finally, *Promises*, a film by Justine Shapiro, B. Z. Goldberg, and Carlos Bolado, follows the lives of seven Israeli and Palestinian children filmed over four years as they develop a sense of who they are and who they are supposed to become. Its sequel, *Promises: Four Years On*, is a sobering return to the children's lives and a study of what happens to children of conflict as they grow up.

Of course, donating money to the organizations listed above is an essential way to help, as is learning about these issues and speaking out to policy makers to create more effective global responses to humanitarian crises and ensuring that children's concerns are not overlooked by those in power. You can organize reading and discussion groups, write to your local paper and to your representatives in Congress, and organize community support for refugees living in this country. The only real solution to these problems, however, is for each and every one of us to work towards a society where opportunities abound for every child and where the next generation will inherit peace as their birthright.

ACKNOWLEDGMENTS

One doesn't write a book like this alone. Since the project began, I have benefited from the kindness, wisdom, patience, and generosity of countless individuals and organizations, some of whom I would like to thank here, in what paltry way these end papers can. I assure the reader, the best parts of this book I owe to the following people and organizations, while any and all errors are entirely my own.

Travel and research were made possible by the generous financial support of Jay Gouline, the Harry and Jeanette Weinberg Foundation, and Ed Vinson and the Mills Corporation.

The staff of Refugees International has been unbelievably helpful in every step of this project. Ken Bacon's initial faith in the idea and in me made the entire endeavor possible. Without the amazing support of the development team over the years—Michelle Kucerack, Antonia Blackwood, Scott Shirmer, and Haida McGovern—I never would have gone on that first mission. It is, however, the advocates who took me into the field with them to whom I owe my greatest debt. Their hard work and patience in the face of surly militiamen, lazy bureaucrats, foreign bacteria, traumas real and imagined, and the odd volcano, as well as their sharp insights and amazing sensitivity to affected populations were a model and an inspiration. I would particularly like to thank Veronika Martin, Joel Charny, and the tireless, courageous, and wise Anne Edgerton, who also devoted a great deal of time and consideration to this manuscript. My debt to board member Jan Weil cannot be put into words for her role as a mentor and as a friend.

I owe all of my interviews and in more than one case, my life, to the many guides and translators I had along the way: Augustin, Eric, Nelson, Daisy, Mohammed, Arber, Ilir, Dada, Florim, and Simon. Additionally, countless staff members from UNHCR and various non-governmental organizations lent me their time and resources in the field. The Women's Commission, Christian Outreach and Development, Doctors Without Borders, and Save the Children were particularly helpful.

I am grateful to Sterling Lord, who saw fit to bring this project into his esteemed literary agency and to the amazing agent extraordinaire, Robert Guinsler, a true friend who believed in me and in this work even when I thought it impossible. Without Robert's guidance, patience, and tenacity, through times when lesser agents would have cut and run, the stories of these young people would never have seen the light of day. I thank you.

Sarah Durand's editorial guidance added immensely to the strength of this book and her insight and hard work on its behalf was an inspiration. Her assistant, Emily Krump, made the whole process completely painless, and for that I am in her debt. I am also grateful to Ilene Smith and Professor Leslie Woodard for gracing me with their valuable time and insightful comments. Additionally, Robert Coles has been an inspiration to this project from a very early stage, putting his faith in an unknown twenty-three-year-old, imparting a fraction of his tremendous wisdom, and lending his effort with amazing humility.

For the initial spark that sent me on my way to that first soccer game in the dirt, I want to thank Julia Hart. And for endless writerly advice, I am eternally indebted to James V. Hart. My dear friends Clea, Alan, Amanda, Brett, Matty, and Michelle listened patiently and encouraged me constantly as this project evolved. Natalie Robin acted as my first line of defense on draft after draft of the manuscript and Victor D'Avella pushed me from day one. James Jayo lit a fire under me and made me believe that indeed people did want to hear these stories.

I cannot give enough thanks to my parents, Anne and Andy, and to my sister, Mandy, for putting up with countless sleepless nights while I was abroad. I am sure they would have preferred I stay home where there were no foreign bacteria, surly militiamen, or active volcanoes. In spite of their fears, they always encouraged me to keep going. And of course, Tim Jones, my North, my South, my East, my West, who quite simply makes all things possible.

Lastly, I cannot name all the children to whom I spoke, but I owe them the greatest debt of gratitude. They gave me their time, their stories, their energy, and their friendship; they shared their fears, their hopes, and the fabric of their lives with a stranger. I pray that one day they will have a chance to read this book and may see the best of themselves in its pages and can, perhaps, tell their own stories, to fill in where my words have inevitably fallen short. They each remain, as ever, in my thoughts. It is for them this work began and it is with them that it ends.

About the author

About the book

Insights,
Interviews
& More . . .

Read on

Meet Charles London

> 66 This created in me a desire to prove [teachers] wrong, a wholly self-centered urge to prove myself, prove that there was a point to my surviving that incubator. 99

I WAS BORN IN BALTIMORE about three months premature. I wasn't supposed to survive, and I lived my first weeks in an incubator. I can't imagine the helplessness my parents felt at seeing this, unable to protect me, doubtful of my survival. My father, an obstetrician, knew all too well that the odds of my survival were low. But survive I did.

Growing up, I didn't excel in sports or school; though I was better at the latter than the former, I was still pretty middle-of-the-pack. Not many teachers thought I would amount to much. This created in me a desire to prove them wrong, a wholly self-centered urge to prove myself, prove that there was a point to my surviving that incubator.

As an undergrad at Columbia University, I studied philosophy but found my real passion in the creative writing department. One of my professors, Leslie Woodard, an inspiring teacher, pushed me to focus and to take my writing far more seriously than I ever had before. She taught me discipline, and gave me confidence in doses just big enough to keep me going—not bigger. I became a fan of the humbling process of the writing workshop.

During my junior year of college, after a year as an intern at *Rolling Stone* magazine focusing on Britney and the ever-shifting ground of reality TV stars, I felt I needed to do something more. That old nagging to prove something came back to me and I decided that writing was the only avenue I had, the only place I might stand out. A friend put the idea in my head to write a book that would benefit a charity I cared about. I had been thinking a lot about the low-scale wars raging around the world at the time, largely outside of the American public's consciousness. This was before 9/11, and there were civil wars in Liberia, Sierra Leone, Sudan, the Democratic Republic of Congo, Colombia, Sri Lanka, and Burma, just to name a few that caught my attention. I began speaking to Refugees International, whose work I had heard of through my mother, about how I could help. But still, in the back of my mind, this remained a way to prove something; it remained about me justifying my survival.

And then I found myself standing in a refugee processing center in Tanzania one sunny afternoon the summer before my ▶

66 And then I found myself standing in a refugee processing center in Tanzania one sunny afternoon the summer before my senior year. 99

senior year. I had never been to Africa before, never experienced war or violence, never really even known great loss. I met men and women and children that day, and in many more missions over the next five years, who survived terrible conditions and awful deprivations every day, often with great humor and hope, always with ingenuity, sometimes with cruelty, but never passively, never the way I conceived of victims. The children most of all, the ones who survived, did so with an amazing amount of effort and creativity. They were not wide-eyed vehicles for suffering or pity as I had seen on the news. It was in getting to know them that I knew I had to write this book, that I had to tell their story as truthfully as I could, that I had to introduce them to people they would never know, and I knew it was no longer about me. It was not about what I had to gain or to prove. The project could no longer be a way to validate my life, but rather became a way to celebrate theirs. Over the next five years of travel and odd jobs to pay the bills, I met hundreds of children who took the time to play with me and to tell me their stories and to whom I owe the biggest success of my young life (writing this book), and the biggest failure of my young life (being unable to fit it all into this book, being unable to keep myself out of the way of the telling). James Agee wrote, in *Let Us Now Praise Famous Men*, "If I could do it, I'd do no writing at all here . . . the rest would be fragments of cloth, bits of cotton, lumps of earth . . . a piece of the body torn out by the roots might be more to the point." I couldn't agree more. ✑

A Conversation with Charles London

When you decided to start your research for Refugees International were your friends and family supportive? Is there one place in particular that they were worried about you traveling to? Where were you the most excited and the most nervous to go?

My first trip was actually planned for Afghanistan—this is back when the Taliban were still in power, the spring of 2001. My parents didn't know much about that area at the time—not many people did—but what they knew they didn't like. I sat down with them and explained the conflict and what I knew of the history—Ahmed Rashid's book, *Taliban,* was helpful—and they kind of relaxed, actually, knowing that I was at least doing my homework before I went, as if knowing the relationship between the Taliban and Unocal oil would keep me safe. But funding didn't come through in time, and the logistics of the mission were too burdensome. For example, it would have been basically impossible for me to talk to girls during my visit and very challenging to do drawings with the children because of the restrictions on artistic depiction under the Taliban. Over the next few years, I intended to go to Afghanistan, but never got the chance. A side effect of this was that when the invasion of Afghanistan began in October 2001, my parents and friends found themselves oddly more informed about the region than most people they knew because of all my talking about the trip I'd planned to take.

Otherwise, they were very supportive. I tried not to tell them too much ▶

5

A Conversation with Charles London
(continued)

beforehand about the specific areas I was visiting or what I'd be doing, lest they go crazy on Google and make themselves sick. It's easy to find bad news about most places in the world, and very easy to imagine the world as a more dangerous place than it is.

Did your research ever feel like it was too much for you to handle? How did you cope with the situations you were faced with? Did you ever feel as though you needed to find a different avenue to help these children, maybe one that involved less travel?

Yes, it definitely got to be too much at times. Hearing these stories and going through the intensity of spending time in some of these places could be emotionally exhausting, but juxtaposing that with college—I spent winter break of my senior year in the eastern Congo—was really hard. I found myself imagining a lot of horrible things, memories would flood back in at inopportune times: one boy's face, scarred from a machete attack; the sound of gunfire in the distance; or just knowing that while I sat in class or in my comfy apartment, the children I had just met were suffering terribly. It wore me out, but—and this might sound trite, I apologize—I had the easy job. All I had to do was tell their stories. They had to survive their stories, live their lives. I got to get on a plane and leave. I figured that the least I could do was keep going, keep working, and do the best I could to highlight their situation. I often wondered what good it would do, what good it could possibly do. Even now I wonder. I hope that knowing these children will enable readers

66 I tried not to tell [my parents] too much beforehand about the specific areas I was visiting or what I'd be doing, lest they go crazy on Google and make themselves sick. 99

to get active, to do something to put an end to the violence in any number of the world's current conflicts or, at least, to give to one of the many organizations that tries to help young people suffering due to these conflicts. But ultimately, the memories are still there; they don't go away; you don't forget these people or places with the passage of time and that can be hard.

I remember after one trip, after I'd been sick with dysentery and there'd been violence in the refugee camp every night, I told my father about some of these things, and he just looked at me and asked if it was worth it, doing this "research." Thinking about the toll it took on me, I wasn't sure how to answer. But I would do it again in a heartbeat, so I suppose it was worth it. Unlike most aid workers, journalists, or scholars, I was there simply to spend time with children. I got to play a lot of soccer and be amazed by them. So all in all, it was a truly positive experience, with moments of terror thrown in from time to time.

What were some of the questions that the children asked you about your life in the United States? Did these change depending on where you were?

The questions were similar around the world. They asked about my family. They asked about New York City—if the buildings were really as tall as they looked in movies. They asked about famous people— usually rappers. In Africa they asked ▶

> ❝ After one trip, after I'd been sick with dysentery and there'd been violence in the refugee camp every night, I told my father about some of these things, and he just looked at me and asked if it was worth it, doing this 'research.' ❞

Siha dreams of being a pilot

Courtesy of the author

A Conversation with Charles London
(continued)

A young refugee's vision of the hardship of war

about Tupac, and they joked about Eminem because I'm white. On the Thai–Burmese border they liked rock music a little more, but still there was curiosity about rappers in America. My friends in the Balkans were really into heavy metal, and they wanted to know about the bands that were new then, that they couldn't get in Kosovo yet. Otherwise, they always wanted to know the names of my family. I spent most of my time answering questions about my sister and my parents. Family was something tangible, something they knew. America was far more of an abstraction for them.

How did the kids react to your recording the conversations? Did they understand that they might one day be part of a book in the United States?

I always explained what I was doing, that I wanted to use their pictures and their stories in a book but I also always asked their permission. More than one said no and those children's stories and drawings, obviously, have not been included. Most, however, were glad to share their stories and many were excited at the prospect of being in a book; even if they couldn't read, they knew that important people got in books and, therefore, they must be important. Some of them helped me pick their pseudonyms that I would use. I didn't always record the conversations, sometimes because of the sensitivity of the material, sometimes because they occurred while walking or playing soccer. The tape recorder and even my pad and pen, made some children clam up. The natural flow of a conversation was often frozen by recording, so I had to scramble to write things down afterward. The idea of pulling out a tape recorder or scribbling furiously as a twelve-year-old told me about being raped or watching his or her parents die just seemed crazy to me. I did, however, record some of the conversations and took notes on far more. It was always funny when a child was acutely aware of what I was doing—like the boy I call Keto, for example—and would pause to give me time to write or aim his words toward the microphone. Several children were very helpful on that end, wanting to make sure I got their stories down accurately, especially because I was going to be telling them again in a book in America. Mostly, they were eager to share.

Were the children aware that their ability to survive was an accomplishment? Were they conscious of their strengths?

Some were, some were not. It varied greatly between children. Some were extremely proud of what they had done so far, but all of them realized they had a lot of unmet needs, needs they were incapable of meeting for themselves—stability, safety—so there really wasn't room for them to be too conscious of their strength. They'd seen the strong die as well as the weak. It's like ▶

> **❝** The idea of pulling out a tape recorder or scribbling furiously as a twelve-year-old told me about being raped or watching his or her parents die just seemed crazy to me. **❞**

A Conversation with Charles London
(continued)

the Hemingway line—"The world breaks everyone and afterward many are strong at the broken places. But those that will not break it kills. It kills the very good and the very gentle and the very brave impartially." I felt like most of these kids understood that line in their bones and, so, couldn't be too secure in anything, which was the great horror of it. They understood the fragility of it all. That caused a lot of anxiety and depression among some of them but also led others toward religious faith. Others it led to reckless behavior, drinking, drugs, and sex. Everyone handles the stress of war and displacement differently.

You mention children coping with their experiences in many different ways. Did you encounter any children who were not coping? How were your interactions with them different?

Yes, I met many children who were depressed, who were anxious, who were angry or violent. Each interaction with each child was different, even if they were coping well, to my eyes. There was no real way to be sure how any of these children were actually coping. The ones who seemed so strong and resilient to me could have been miserable; I might have caught them on a good week or a good day. Some of them changed so dramatically from day to day that I felt like I was dealing with an entirely different person. That's the challenge of working with children anywhere, but especially with writing about them. The very nature of this project is trying to fix them into words at a time when they are in such great transition. Anything that is true at one moment may not be true the next. Assumptions about how well or how poorly a child was coping were always extremely dangerous for me.

In Kosovo, religion seemed to be a prominent factor in the lives of the children, and it lay behind the prejudice and the war. Was religion a factor in Africa or Asia? Did you see any correlation between the children's faith and their hope for the future? Did the complication of religion affect how they were dealing with their situations?

Religion was a factor everywhere. I found that the children who had religious faith often professed to be more able to deal with terrible things, and many of them claimed it was only their faith that got them through. But children with strong political beliefs also seemed to have the same attitude in relation to their political faith. Generally, though, the children who believed in something had more energy, seemed less prone to torpor and hopelessness. This was true of Muslims, Jews, Buddhists, and Christians. But it was not universally true. I met children who didn't claim to have any religious or political faith, yet still appeared to be doing well, taking care of others, surviving, and staying happy. The factors that lead to resilience in a child affected by violence are complicated and no one—certainly not me—knows for sure what role things like religion play. Hope, either in a just afterlife or a better mortal life, pretty much always helps though. ❧

66 Assumptions about how well or how poorly a child was coping were always extremely dangerous for me. 99

11

Outtake
The Roma of Kosovo

ONE GROUP OF CHILDREN with whom I spent time, but did not include in the main text, were the Roma of Kosovo. The Roma, also called Gypsies, are the poorest of the ethnic groups in Kosovo, confined to protected enclaves like the Serbs but given the worst land and the least government support. Due to alleged complicity with Serb paramilitary operations during the ethnic cleansing, they are reviled by much of the Albanian population and often blamed for any acts of crime in the areas where they live. As Ibrim, a sixteen-year-old Roma boy living near Pristina, explained it to me, "The Albanians treat us like dogs and the Serbs treat us like pet dogs." Ibrim had the dark skin and bright green eyes typical of the Roma. "Because we come from India," he said.

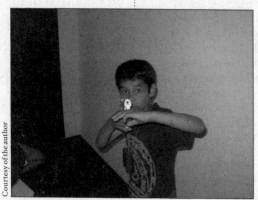

Courtesy of the author

Ibrim, a young Roma boy in a youth center in Kosovo (the Roma, or Gypsies, are the smallest minority group in Kosovo, suffering from the mistrust of both the Serbs and Albanians)

He wanted to be a journalist when he grew up, though his opportunities for schooling were limited and he suffered from the lack of mobility facing all of Kosovo's ethnic minorities. Their neighborhoods or camps were surrounded by barbed wire and guarded by NATO troops. Ibrim did not feel safe in Pristina, and the dirty looks the cab driver gave us when we got in together to drive back to the Roma displacement camp made it clear how welcome this boy was in the city. He looked out the window and snapped pictures with the disposable camera I'd given him. An old Albanian man shuffled by the side of the road. Two young men sat smoking cigarettes by the curb. Their clothes were dirty and the way they leaned on each other made them look drunk. It was eight in the morning, but with unemployment so high, alcoholism rates were high too.

"In America, you have poor people like we do," Ibrim said, trying to make me understand the Roma's place in the social order. "And you also have niggers. We Roma are the niggers of Kosovo."

I balked at his statement and he asked why I was so uncomfortable.

I explained to him how the word "nigger" was used in America, what it meant to hear it and to use it, the history of exploitation, injustice, and bigotry associated with that word.

"Exactly," he said. "Then it is exactly what I wanted to say. Also, I love rap music." He laughed and patted me on the back, fully aware of how uncomfortable he'd made me, which it seems, was exactly his intention.

Nearly ten years after the war, the Roma still find themselves displaced. No other countries want to take them in. Their history in the region is contentious. They are often accused of Nazi collaboration, of collaboration with the attacks on Albanians, of spying, of large-scale looting, and of war crimes. Even when the accusations aren't clear, they are seen as guilty of something, and best avoided by decent people. I was warned more than once not to visit their enclaves or to watch my wallet when I did. People in Kosovo, by and large, would much rather the Roma did not exist. ▶

> **❝** The dirty looks the cab driver gave us when we got in together to drive back to the Roma displacement camp made it clear how welcome this boy was in the city. **❞**

The Roma of Kosovo *(continued)*

As a Refugees International team visiting the region in 2005 discovered, the United Nations Mission in Kosovo (UNMIK) delayed for more than a year the evacuation of hundreds of internally displaced Roma, Ashkali, and Egyptians from highly toxic displacement camps poisoned by lead. As their report indicated, "The UN's intransigence is condemning children to severe lead poisoning, possibly resulting in irreversible mental and physical retardation." I was reminded of the extremely high asthma rates in Harlem and the South Bronx and thought how apt Ibrim's racial critique of both his country and mine had been.

Even now, in my work, the Roma have been removed from the central narrative of Bosnia and Kosovo, relegated to the back of the book. I found it difficult to place their story in the story of the conflict as I was telling it. The Roma were not on the minds of either the Albanian or Serb children, though both those groups talked about each other all the time. No one, it seemed, gave much thought to the Roma, but the Roma.

"We just want to be left alone," Ibrim said. "But people are always chasing us around with guns, accusing us of things. Sure we have criminals and some Roma did the things the Albanians say we did, but only for survival, only because the Serbs forced them to. I was little during the Milošević war, but if I'd been older, I wouldn't have hurt anyone, unless they attacked me first. The Serbs and the Albanians will fight until they're all dead, you know? They both want Kosovo for themselves. But we Roma will still be around." ∽

Letter from a
Lost Girl of Sudan

(1)

THE LOST GIRLS
GENERAL LIVES HISTORY
GIRLS ORPHANED AND THROWING OUT
OF HOME

It was in the year of 1987 when mass killing of people,
the destruction of properties and burning of our villages
by people whom most of us have witnessed to be
wearing uniform which were green in colour and with
brown skin surely seen by most of us who their
areas were destroyed at daytime by enemies.
Due to that war, some of us who God, saved their live
escaped narrowly in the war and flew to Ethiopia for lives safety
On the way to Ethiopia, we were faced with a lot
of difficulties like:- hunger, thirsty, many attacks
on the way by hostile tribes and tireness for always
walking all the time i.e. day and night in the jungle with
bare foot untill we reached Panyudo Refugee camp
in Ethiopia after losting many lives of our colleagues
on the way.

Courtesy of the author

15

Refugees International

REFUGEES INTERNATIONAL was founded in 1979 to advocate for U.S. protection for Vietnamese boat people, refugees from the killing fields of Cambodia, and Laotian freedom fighters. We played a key role in convincing President Jimmy Carter to boost refugee admissions from Southeast Asia. Since then Refugees International has worked throughout the world to increase resources and protections for more than 33 million refugees and displaced people, serving as a powerful voice for lifesaving action.

Our advocacy, based on up-to-date, independent field assessments of refugee conditions, gets results. Each year Refugees International conducts twenty to twenty-five field missions to assess crisis situations that have caused displacement. Refugees International's advocates spend several weeks in the field interviewing displaced people, government officials, and the staff of agencies that are responding to the crisis. On each mission—along with recording the need for services such as food, water, shelter, and protection from harm—advocates explore a variety of other issues, such as health services, access to education, human rights abuses, and circumstances specific to women and children.

Refugees International does not accept government or UN funding and relies solely upon the generosity of informed and engaged organizations, corporations, and individuals. ∾

Visit www.refugeesinternational.org to donate and to find out other ways you can join our lifesaving advocacy.

66 Refugees International does not accept government or UN funding and relies solely upon the generosity of informed and engaged organizations, corporations, and individuals. 99

Don't miss the next book by your favorite author. Sign up now for AuthorTracker by visiting www.AuthorTracker.com.